Inside the Rice Bowl

Agrarian Crisis and the Rising Distress of Farmers in West Bengal

Barun Kumar Paul
Kunal Chakraborty
Gopa Samanta

RAJESH PUBLICATIONS
NEW DELHI - 110002

Inside the Rice Bowl

Agrarian Crisis and the Rising Distress of Farmers in West Bengal

Published by :
Sanjay Gupta
RAJESH PUBLICATIONS
1, Ansari Road, Darya Ganj, New Delhi - 110002
Ph.: 011-43084725
Website : www.rajeshpublications.com
Email : rajeshpublications@gmail.com

First Published : 2023

ISBN : 978-93-91139-37-7

Printed in India by :
Belwel Process
Delhi - 110002

We dedicate this book to our fathers and grandfathers who raised crops in their own small farms located in Purba Bardhaman, Birbhum and Bankura districts of West Bengal to ensure our families' food security.

Acknowledgement

We are thankful to the people of our study sites for participating in our research process, for their valuable suggestions and comments, and also for their kind help in many other forms such as providing water and food, and giving space for conversation and discussion with them within their household premises. We would like to thank Asish Nayek, ADA (Socio-Economic & Evaluation), Burdwan, who helped us to understand cost calculations and rendered valuable information whenever sought for. We wish to express our thanks to the different institutions from which we received help during our field survey. We extend our thanks to the officials of Galsi-I Agricultural Development Office, especially to Ujjwal Mandal, Krishnen Murmu, and Arjun Chandra Soren for their kind help. We acknowledge Asit Santara, Ashok Mondal, Suman Ghosh, Swarup Mondal, Jibon Ruidas, Baidyanath Barik, Dr. Tapas Kumar Chakraborti, Avik Kumar Roy, and Santosh Malik for their assistance during the field survey. We are thankful to the faculties of the Department of Geography, The University of Burdwan, and of many other institutions for their valuable suggestions while presenting our research findings. We sincerely acknowledge the editing services of Ankita Kaul, who meticulously read, edited, and commented on the manuscript.

Preface

There is a general consensus that Indian agriculture is in a crisis phase. This agrarian crisis has been described from different perspectives by different disciplines. However, there is a lack of a holistic study on agriculture and on farmers, to rightly understand the depth of this agrarian crisis that persists in West Bengal and in other parts of India. Against this backdrop, the present research is an in-depth attempt to address the ongoing agrarian crisis using a multidisciplinary approach to understand the complex situation of agriculture as well as of farmers. Though the study is based on only two CD Blocks of Purba Bardhaman district of West Bengal, it tries to provide a broad picture of farmers and farming in the State.

This book focuses on the agrarian crisis, and considers crisis in terms of both the process of farming and the farmer. Here, the unit of study is the 'farmer household', where farmer households have been classified into six categories to present the inter-class variations on various aspects.

The entire book has been divided into nine chapters. The ***first chapter*** introduces the research. In this chapter, the central research question, objectives and significance of the study, methodology, and a brief description of the study area have been presented. Besides, a literature review has been included in order to identify the relevant research gap.

The ***second chapter*** focuses on agricultural growth in West Bengal in general and in Purba Bardhaman district in particular during different periods. It explores how farmers were in a distressed situation even in the historical past. Using NSSO data on Situation Assessment Survey, this chapter then outlines the current pitiable conditions of farmers in West Bengal.

The ***third chapter*** explores various issues related to major input use in cultivation, with special emphasis on paddy cultivation. This chapter introduces the past (before 1995) and present scenario of input use, such as seed, fertilizer, irrigation, human labour, and farm implements, thereby highlighting various problems related to input use that have arisen over time.

Costs and returns from paddy, mustard, and potato cultivation have been explained in ***chapter four***. As paddy is the main crop of the study area as well as in West Bengal, more emphasis has been given on profitability of paddy cultivation. This chapter tries to answer two important questions: 1) Is cultivation profitable enough for the farmers? 2) Why are farmers sticking to crop cultivation, especially paddy cultivation?

The ***fifth chapter*** explores the limitations in the government support system meant for agricultural development and farmer welfare. It is evident that this support does not equally reach all categories of farmers like the landless, marginal or small. Rather, affluent farmers benefit more from these schemes. The findings of this chapter are exclusively based on empirical survey.

The ***sixth chapter*** explains how tractor use has created problems in spite of having many advantages; how it is responsible for growing chemical fertilizer use and declining bio-fertilizer use. This chapter shows that tractor use is an obstacle for achieving sustainability in agriculture. Besides, this chapter has highlighted some real issues which prevent farmers from adopting modern technologies in paddy cultivation.

Chapter seven looks into the changing lifestyle and increasing distress of the farmer community. This chapter examines how social change led by modernization and consumerism is exerting financial pressure during the ongoing agrarian crisis and job crisis. This chapter tries to establish that the agrarian crisis is not an isolated matter; rather, it is related to the economic, social, and political environment of the country, and thus it creates a complex scenario.

The ***eighth chapter*** investigates the growing system of short-term tenancy. This chapter discusses the characteristics of this tenancy system, the causes for its growth, and thereafter analyzes how this tenancy system is creating an acute distress situation among the poor farmers.

Chapter nine includes major findings and contributions of the research. Besides, this chapter argues on the multi-dimensional nature of the existing agrarian crisis triggered by an insufficient government support mechanism, Green Revolution led agrarian crisis, limitations behind procurement of paddy at minimum support price (MSP), declining household savings with the spread of consumerism and economic liberalization, rise of input costs, joblessness and forced migration of agricultural labourers, etc. In spite of several issues and growing agrarian crisis, most of the mainstream political parties either remain disinterested or suppress the problems. Finally, this chapter ends by questioning the policy shift towards multi-pronged state-sponsored strategies for achieving sustainability in agricultural development.

Contents

List of Tables

List of Figures

Acronyms and Abbreviations

AoA	: Agreement on Agriculture
ADA	: Assistant Director of Agriculture
ANOVA	: Analysis of Variance
APL	: Above Poverty Line
ATM	: Assistant Technology Manager
ATMA	: Agriculture Technology Management Agency
BFBY	: Bangla Fasal Bima Yojana
BGREI	: Bringing Green Revolution to Eastern India
BPL	: Below Poverty Line
BTM	: Block Technology Manager
CACP	: Commission of Agricultural Costs and Prices
CD	: Community Development
CHC	: Custom Hiring Centre
CoC	: Cost of Cultivation
CoP	: Cost of Production
CPIS	: Coconut Palm Insurance Scheme
CPC	: Centralized Procurement Centre
CPIAL	: Consumer Price Index for Agricultural Labourers
CRSMF	: Crop Research and Seed Multiplication Farm
CSDS	: Centre for Study of Developing Societies
DAP	: Di-Ammonium Phosphate
DVC	: Damodar Valley Corporation
FAO	: Food and Agriculture Organization
FRT	: Fixed Rent Tenancy
GDP	: Gross Domestic Product
GP	: Gram Panchayats
HYV	: High Yielding Variety
IADP	: Intensive Agricultural District Programme
IIFFCO	: Indian Farmers Fertiliser Cooperative Limited
JLG	: Joint Liability Group

KCC	:	Kisan Credit Card scheme
KPS	:	Krishi Prayukti Sahayak
KVK	:	Krishi Vigyan Kendra
LPG	:	Liquefied Petroleum Gas
MGNREGS	:	Mahatma Gandhi National Rural Employment Guarantee Scheme
MPCE	:	Monthly Per capita Consumer Expenditure
MSP	:	Minimum Support Price
NAIS	:	National Agricultural Insurance Scheme
NAMET	:	National Mission on Agricultural Extension and Technology
NFSM	:	National Food Security Mission
NITI Aayog	:	National Institution for Transforming India
NPK	:	Nitrogen Phosphate Potash
NSSO	:	National Sample Survey Organization
OBC	:	Other Backward Class
PACS	:	Primary Agricultural Credit Societies
PDS	:	Public Distribution System
PMFBY	:	Pradhan Mantri Fasal Bima Yojana
PSCA	:	Parliamentary Standing Committee on Agriculture
SAS	:	Situation Assessment Survey
SC	:	Schedule Caste
SHC	:	Soil Health Card
SHG	:	Self Help Group
SRI	:	System of Rice Intensification
SRR	:	Seed Replacement Ratio
ST	:	Schedule Tribe
TL	:	Truthfully Labelled
WBCIS	:	Weather Based Crop Insurance Scheme
WTO	:	World Trade Organization

1 Introducing the Agrarian Crisis in India

Putting the Study in Perspective

Since the neo-liberalization of the Indian economy in the 1990s, the contemporary agrarian crisis has witnessed many changes and transitions. The reforms led by liberalization have led to a series of structural changes and consequently the crises have deepened in Indian agriculture. The current crisis in Indian agriculture incorporates the dimensions of 'agricultural', i.e., the developmental crisis and of 'agrarian', i.e., the livelihood crisis of the farmers (Mishra & Reddy, 2011). The agrarian crises are manifold, such as slow growth of agriculture and less GDP share, stagnation of rural employment in agriculture (Mishra, 2013; NSSO, 2003), low level of agricultural productivity (Lerche, 2011), food security issue, farmers' suicides (Reddy & Mishra, 2010), technology, functioning of credit and capital investment to farmers, declining crop prices and so on (Siddiqui, 2015). However, these issues remain neglected, with very little attention being paid to those in Indian economic growth discourses.

Farmers, the sole suppliers of food to millions of people in this country, remain ignored and are literally pushed to distressed conditions (Reddy & Mishra, 2008; Bhoi & Dadhich, 2019). As per the Situation Assessment Survey of Farmers, 2003, about 48.6 percent farmer households were indebted, and about 40 percent of the farmers all over India did not like agriculture as a profession. This situation had not changed even after a decade. As per the Situation

Assessment Survey of Agricultural Households of 2013, about 51.94 percent agricultural households were indebted, and about 36.4 percent agricultural households possessed a below poverty line (BPL) ration card. In another study conducted by the Centre for Study of Developing Societies (CSDS) in 2018, about 76 percent farmers opined that they would prefer to do some other work rather than farming (p.11). Though the results of these studies are not comparable, it is quite clear that the acceptability of agriculture as an occupation to the rural people is declining day by day.

The process of transformation of the agricultural workforce or depeasantization is either growth-induced or distress-driven (Singh et al., 2009). A major cause of not opting for agriculture as an occupation is unprofitability (NSSO 59th Round, 2005; CSDS report, 2018, p.10). According to Swaminathan (2016), the father of the Green Revolution in India, there is no other profession which has as low returns as agriculture. In fact, even aside from low returns there are various problems in Indian agriculture. Some of these problems like over-dependence on monsoon, small landholdings, high rate of dependence on agriculture, etc., were present even before independence. Though the Green Revolution has increased food grain production, it has further created some social and ecological problems, such as social inequality, soil degradation, declining groundwater, air pollution, etc. After-effects of the Green Revolution are serious concerns and major causes of the present agrarian crisis (Brown, 2018, p.2).

In addition, economic liberalization during the 1990s aggravated the misery of Indian farmers. Economic liberalization had a great impact on the decline in public investment in agriculture and slowdown in agriculture (Bhalla & Singh, 2009). The post-economic reform period is also characterized by rising input costs of cultivation, low food prices, and growing indebtedness among farmers. A huge number of farmer suicides have been reported across India during this period (Nagaraj et al., 2014).

All these agricultural problems have increased over time, and have become more complicated in the present era of 'jobless growth'.[1,2] The 55th round of NSSO data reveals that 40 percent of the rural incomes are coming from diversified and non-farm sources. Though the per-capita land holding is decreasing, the number of small and marginal farmers is on an increasing trend. With the decline of land and labour productivity, agricultural unemployment is forcing the rural labour force to migrate elsewhere for non-farm work (Murthy, 2013). The contribution of agriculture to the national gross value added (GVA) has decreased from 15 percent in 2015-16 to 14.4 percent in 2018-19 (Economic Survey, 2018-19), though about 70 percent of the rural households still depend primarily on agriculture for livelihoods (Food and Agriculture Organization, n.d.). However, according to the State of India's Livelihood Report (2019), agricultural GDP has been declining since 2016-17. As the share of agriculture in National Gross Domestic Product (GDP) is declining day by day, industry and service sectors should absorb the surplus labour and create job opportunities. However, these sectors have failed to generate sufficient job opportunities (Thomas, 2008; Tajani, 2015). It is feared that more people in India than needed will have to stay on the farm, or keep returning to it after brief stints as short term contract workers on non-farm projects, as long as there are no alternatives (Jagannathan, 2018, p.8), in spite of declining GDP from agriculture.

In this situation, the government is trying to help farmers in many ways. Various schemes for agricultural development have been introduced. However, these facilities are still inadequate for agricultural development, as well as for farmers' welfare. In an interview, Sharma (2017), agricultural scientist and activist, opined that the budget of MGNREGA is higher than the budget for agriculture, and about 75 percent of the budget allotment for agriculture goes to the.

1. Nadeem (2019) opined India is experiencing jobless growth since 2004-05.
2. Economic growth with falling employment elasticity is termed jobless growth.

industries. Sharma further opined that farmers who cultivate cash crops like tomato or potato are also not getting the right price for their produce due to lack of proper policy and 'mismanagement'.

Over the past few years, many newspaper reports have been published on agriculture-related issues, deepening farm crisis, and on farmers' protests. Like other parts of India, farmers of West Bengal are also facing problems. As per the CSDS report, 2018, the assessment of the general condition of farmers is most negative in West Bengal (p.21). Now-a-days, the linkages between agrarian distress and resultant out-migration are quite familiar in many parts of West Bengal (Thapa & Mishra, 2022). Incidents of farmer suicides have also been reported. Therefore, there is a need for a holistic study on agriculture, as well as on farmers to rightly understand the depth of the agrarian crisis persisting in West Bengal and in other parts of the country. In this context, this study attempts to understand the multi-dimensional nature and depth of the agrarian crisis of the paddy farmers of West Bengal in general and of the Purba Bardhaman district in particular.

Changing Agrarian Questions in Post-Independence India

Indian agriculture has undergone several changes and developments since independence. Indian economy is primarily based on agriculture, which contributes a significant proportion of the national GDP. Although 60 percent of the total workforce is still engaged in the agricultural sector, after sixty years of independence the income from agriculture has started declining from 50 percent in the 1950s to 18 percent in 2007-08 (Tripathi & Prasad, 2009). After independence, these changes and transformations can be traced out in four phases: from 1947 to the mid-60s, from mid-60s to the 1980s, from the 1980s to 1991, and the last phase from 1991-92 onwards.

The pre-Green Revolution phase (1950-1966) was marked by a sharp decline in agricultural GDP, i.e. from 2.78 percent to 1.06 percent. This period had witnessed institutional changes, agrarian reforms, construction of major irrigation projects, and proliferation of cooperative credit institutions to give access to credit to the farmers. The major agrarian reforms introduced during this phase incorporated tenancy reforms, Zamindari Abolition Act, abolition of intermediaries in land holding, and fixing the ceiling of landholding (Arora, 2013). Land reforms introduced in the first decades after independence suggested that it has a differential impact on both the large and small tenants. The large tenants benefitted more, while small tenants were adversely affected. In between 1951 to 1966, there was a decline in tenant-operated areas leased across different states of India. This new agrarian reform increased the importance of commercial tenancy, which was more exploitative than the earlier feudalistic type of tenancy. It favoured the erstwhile intermediate class or medium landlords (Joshi, 1974). Incentive price policy popularly known as Minimum Support Price (MSP) was introduced in due course by the Agricultural Price Commission, which encouraged the farmers to use modern technology in agriculture.

The advent of the Green Revolution (1966-1980s) incorporated the adoption of new agricultural technology with high yielding variety (HYV) of crops, multiple cropping, expansion of irrigation facilities, modern farming with increased use of fertilizers and agro-chemicals, price support and marketing. In this phase, the production of food grains increased from 74 million tonnes in 1966 to 105 million tonnes in 1971-72. Declining rural poverty (64% in 1967 to 56% in 1973) was also noticed in this period (Banka et al., 2021). Another major breakthrough was nationalization of major commercial banks for enhancing credit flow in the agricultural sector. The most significant aspect of the Green Revolution was the achievement of self sufficiency in food grain production. The first phase in the pre-Green Revolution period was characterized by deficit food production. But during the second phase, the post-Green Revolution period, i.e., from the mid-1960s to the 1980s, the country achieved self-sufficiency in food grain production. During the 1970s, inter-crop and inter-regional imbalances in crop yield occurred, and during the 1980s there was a shift from cereal to non-cereal and from crop to non-crop sector (Radhakrishna, 2002).

During the era of pre-economic reform (1980s to 1991), government subsidies and support programmes in the agricultural sector increased considerably. As a result, the growth of non-food grain output accelerated the agricultural GDP in the Indian economy. While the public sector investment in agriculture declined, farmers' investment in agricultural infrastructure showed an increasing trend. Together with these, the accelerating commercialization of agriculture which has been taking place since the 1980s is reflected in the diversification towards non-food grains, increased proportion of market arrivals of output, and dependence on purchased inputs. During this phase, enhanced food grain procurement, increased literacy among farmers, improvement in marketing and other infrastructure, etc., have helped greatly in agricultural commercialization. However, commercialization has made farmers more vulnerable to market forces and market risk, and this vulnerability is aggravated by the presence of monopolistic elements in the market (Nadkarni, 1996).

The Current Agrarian Crisis in Neo-liberal India

Indian agriculture, in the realm of neo-liberal economic reforms which started from the 1990s, has witnessed many kinds of distress in the country's agrarian economy. The neo-liberal agrarian reform deals with three major aspects: deregulation, privatization, and globalization. With a sharp decline in the growth rate of agricultural output, it has intensified many agrarian crises across space, time, and among different classes in rural India. This has led to expenditure reduction and rise of input costs, removal of restrictions on agricultural product import and export, reduction in the rate of institutional credits, reduction in rural development expenditure and enhanced penetration of foreign capital in agriculture (Athreya, 2013; Arora, 2013). During the reform period many scholars were of the view that low crop productivity was the result of low input use, which is caused by a poor network of input distribution, lack of resources and non-availability of adequate credit for purchasing modern inputs, and stagnation in public investment in irrigation and other infrastructure. Investment in irrigation, increased power supply, more fertilizers, competitive seed market, improved terms of trade for agriculture, and measures to mitigate risk are the main requirements for the revival of agricultural growth.

The small and marginal farmers and the agricultural labourers who constitute the majority of the agricultural workforce were the most adversely affected. The neo-liberal reforms largely challenged the wage and employment status and also questioned the well-being of the agricultural labourers. According to the NSS data, the number of agricultural labourers increased during the 1990s (Jha, 2007). Moreover, the share of agriculture in total employment which was 68 percent in 1983-84 declined to 58 percent in 2006-07 (Das, 2012). According to the Situation Assessment Survey conducted by the NSSO in 2003, about 96.2 percent of the farmer households owned less than 4 hectares of land. Between 1987-88 to 2004-05, there has been a significant rise in landlessness among the agricultural households. From 1990 to 2004-05, Indian agriculture witnessed deceleration in growth of output. In consequence, squeezing of income and employment opportunities in the farm sector increased the indebtedness among farmer households up to 48.6 percent in 2003 (Pillai, 2007).

It was Radhakrishna (2002) who further opined that shifting of the workforce from farm sector to non-farm sector is a normal process of development, but the major concerns of the post-reform period are the very slow pace of this process, the high rate of unemployment among educated rural youth, increase in casual employment rather than self-employment or regular salaried employment, increase in real wages in spite of slowdown in agricultural growth, shifting of non-farm activities from rural to urban areas, and widened rural-urban disparities. Moreover, liberalization in the agricultural sector has led to concentration of land in a few hands, which has perpetuated small peasantry and proletarianization (Haque, 1996). Studies find that increasing concentration of small and marginal farmers in agriculture will induce semi-proletarianization, which may displace many people from agriculture and thus increase informalization in the employment sector (Yadu & Satheesha, 2016). In the post-reform period, the fall in public sector investment in the agricultural sector argued for inviting private investment in agriculture. Therefore, increasing costs of cultivation, mounting pressure of indebtedness, negative impact of economic liberalization, and low returns from agriculture are the prime concerns of agrarian distress in India (Kumar, 2020).

In Indian agriculture the term 'agrarian crisis' bears the meaning of not only the deceleration in agricultural growth and productivity, but also the agrarian distress experienced by the farming community. Agrarian crisis is 'structural and institutional in nature'. The multi-dimensional nature of the crisis also includes 'growing marginalization' in agriculture, as well as 'failure in support system'. In Indian agriculture, marginalization is reflected by the predominance of small and marginal farmers in operational land holding. In spite of land reforms, small and marginal farmers accounted for 86 percent of land holdings and the pattern of land distribution seems to be skewed (Reddy & Mishra, 2010). During the reform period, the failure of support systems and changing policy interventions brought many hardships to the small and marginal farmers. The introduction of new technology did not help the small and marginal farmers much due to their lack of access to resources. However, Mohanty (1999) has argued that in spite of removing inequalities, agricultural modernization often increases disparities in many places. Therefore, the small and marginal farmers face many disadvantages and deprivations in terms of accessing credit, insurance, and low returns from agriculture.

High input cost is one of the major causes of the agrarian crisis in India. In the era of new economic reforms, the unregulated market has caused a tremendous increase in the price of seed. In comparison to the investment in seeds, farmers get less returns and consequently they are pushed into indebtedness (Ahoy, 2016). According to the 59th round of NSS report in 2005, about 50 percent of the farmer households were indebted (Sharma, 2019) and the root cause was the rising agricultural expenses. In 1991, with deepening economic crisis and following the conditions given by IMF (International Monetary Fund), the adoption of the Structural Adjustment Programme (SAP) resulted in the withdrawal of agricultural subsidies, which also increased the input costs (Siddiqui, 2021; Srinivasulu, 2018; Patnaik, 2006). From 1997-98 to 2003-04, the fertilizer subsidy was reduced considerably in the union budget (Reddy & Mishra, 2008). The cutting of the fertilizer subsidy has made agriculture more costly and unprofitable (Aerthayil, 2008).

In the present era, Asian agriculture also faces the challenges of demographic pressure, fragmentation of landholdings, migration, and rising female and ageing farm population. Therefore, planning related to transformation of agricultural small holding is needed to achieve farmers' sustainability (Viswanathan et al., 2012). However, Deokar and Shetty (2014) in their study opined that Indian agriculture is doing better since 2004-05 due to policy initiatives. Increased allocation for various departments related to agricultural development, implementation of the National Horticulture Mission, reformation in the compact and systematic extension system, increased price of agricultural commodities in the international market, and initiation of the Bharat Nirman Project resulted in substantial increase in private investment, which in turn resulted in about 4 percent growth for the period of 2004-05 to 2013-14 compared to 2.4 percent growth during 1994-95 to 2003-04.

In the post-reform period the credit disbursement pattern of most of the banks changed drastically. Loans were not easily available to the farmers at an affordable interest rate.

Credit facilities extended to the farmers were also reduced in the post-reform period. Liberalization policies advocated that banks should function on a profitability basis (Siddiqui, 2015). As a result of this reform the institutional indebtedness for the small and marginal farmers increased upto 68 percent in 2003. Though their share of credit from institutional sources had declined between 1993-94 and 2003-04 (Reddy & Mishra, 2010), large farmers benefited in terms of money lending from the banking sector (Chavan, 2007). Thus, the declining credit facilities from institutional sources forced the small and marginal farmers to depend largely on informal sources. Together with the rising input costs, high commodity price, reduction of public investment in agriculture, and increasing privatization has been pushing the small and marginal farmers to commit suicide in rural India (Mohanty, 2005). The post-reform agrarian distress has resulted in an increasing number of farmer suicides from 1997 to 2021, caused due to farmer indebtedness and high risk in agriculture (Siddiqui, 2015).

In Indian agriculture, the impact of globalization is a debatable discourse. According to Bhattacharyya (2016), the neo-liberal reform 'was not based on economic and political consensus'. In spite of several benefits, Indian agriculture has been affected greatly in the post-liberalization period. With the decline of agricultural GDP growth from 3.3 percent in 1980-95 to 2 percent during 1995-2005, the agricultural sub-sector has recorded a downfall between 1990-97 and 2003-04. Liberalization policy not only reduced the pesticide subsidy but also increased the price of fertilizers by more than 300 percent. The situation eroded further after the inception of the WTO. For reducing import restrictions, the price of cotton has gone down by 60 percent since 1995, resulting in farmer suicides in Maharashtra. The report published by the Planning Commission's Task Force on employment generation depicts a significant reduction in the number of people employed in agriculture.

The growth rate of agricultural employment also

decelerated between 1993-94 and 2004-05 (Shetty & Gowda, 2013). Bhalla (2004) also pointed out that setting up an institutional mechanism for helping a large number of small and marginal farmers, large-scale subsidized imports, non-tariff barriers like sanitary and phyto-sanitary standards imposed on imported agricultural commodities to developed countries, fluctuation in the international market, ensuring food security, etc., are the major critical challenges of globalization in India. In the case of export competition, India is unable to enjoy a comparative advantage. Some powerful lobbies play power politics to achieve favours from WTO negotiations and force other countries to accept it (Ravikanth, 2014). Thus, globalization controls our government policies adversely, in turn creating a distress situation in the agro-economy (Bhukta, 2007).

Agriculture and Environment

There is a consensus that due to the adoption of technology-based agricultural practices, Indian agriculture has faced many environment-related problems like soil degradation, loss of biodiversity, exploitation of groundwater and energy, chemicalization, etc. Climate change and its impact on agriculture is also another environmental issue. Over the last few decades, India has achieved 'self-sufficiency' in food production. However, with increasing demand linked to the growth of population, the strategies to increase food production have heavily exploited the natural resources. The strategies adopted during the Green Revolution era have resulted in widespread groundwater resource depletion and other environmental problems such as loss of soil fertility. Increasing use of chemical fertilizers has also caused groundwater pollution and increased emission of greenhouse gases (Abrol & Sangar, 2006). All these are threatening the sustainability of future agriculture. Therefore scientists, scholars, and researchers are giving more importance to environmental sustainability issues.

Chemical fertilizers are normally used to increase crop productivity. In India, the demand for chemical fertilizer is increasing rapidly. The consumption of nitrogen and phosphorus based fertilizers has increased by about 13 to 14 times from 1970 to 2020 (FAI, 2020). However, over-fertilization may cause acidification of soil, excessive leaching, and removal of soil organic matter. The application of NPK fertilizers in the long run has some negative consequences on soil health (Neog, 2018). Excessive use of agrochemicals, i.e., both chemical fertilizers and pesticides, has an adverse impact on surface and groundwater contamination (Srivastav, 2020).

Pesticides have made a significant contribution in increasing farmers' production and income from agriculture. In India, from 1950 to the late 90s, the indiscriminate use of agro-chemicals such as pesticides caused damage not only to the environment but also to living beings (Rajendran, 2003). According to the FAO, India consumed about 58,160 tonnes of pesticides in 2018 (Nayak & Solanki, 2021). Empirical findings have suggested that pesticide use actually lowered the agricultural productivity and adversely affected human health. Even the cost of crop loss by pests in many instances is less than the cost of pesticide-induced illness (Rola & Pingali, 1993).

Groundwater is one of the prime sources of irrigation in India. With the development of groundwater-based irrigation since the 1970s, it provides nearly 60 percent of the country's irrigation supply (Sibert et al., 2010). India has recorded an expansion of irrigated area from 58 percent in 1961 to 60.4 percent in 2016 (World Bank, 2017). Depletion of groundwater resources due to irrigation has increased globally, to which India contributes about 23 percent of the total loss (Dalin et al., 2017). The aquifers in many parts of India are reaching critical levels due to huge extraction of water (Danger et al., 2021). Scholars have argued that rapid withdrawal of groundwater, on an average, has caused a drop in the water table by more than eight metres since the 1980s (Aeschbach-Hertig & Gleeson 2012; Sekhri 2012; Rodell et al., 2009). Over-exploitation of groundwater, declining soil fertility, increasing intensity of pests and disease, groundwater pollution due to leaching of chemical fertilizers and pesticides, air pollution caused by burning of rice straw, and lack of skill among farmers are the major problems faced by the rice-wheat crop system of Uttar Pradesh, Bihar, and West Bengal (Chand & Haque, 1998). According to Sarkar and Das (2014), groundwater depletion in Punjab for paddy cultivation has been caused by free electricity supply after 1997 and higher MSP of paddy. They also opined that lack of maintenance resulted in drying up of many canal networks in Punjab, and depletion of groundwater resulted in drying up of shallow tube wells in the first phase of the Green Revolution, resulting in the spread of deep tube wells locally called submersibles. Many findings have also argued that depletion of groundwater has not only reduced the production of crops, but in the near future with some amplifying climate change issues it will further stress the agricultural distress in terms of production and cropping intensity (Zaveri et al. 2016, Jain et al. 2021; Zaveri & Lobell, 2019).

Burning of crop residue is another environment-related issue in agriculture as it causes loss of soil fertility as well as large scale air pollution. Chaudhary (2018) opined that since farmers are habituated to killing microbes using chemical fertilizers, so crop residue burning is the next logical step to them. Both Gill (2018) and Chaudhary (2018, p.60) admitted that farmers burn paddy straw as it is less economical for on-farm usage and there is very short time slot between harvesting paddy and sowing of wheat (the next crop). However, Chaudhary ascertained that ignorance of farmers about soil and farm economics are the leading causes for crop residue burning, and on-farm usage of the crop residue through the adoption of organic farming is the best option to solve crop residue burning related problems, rather than off-farm usages.

Farm Income and Profitability

In Indian agriculture, policy issues related to farm income and profitability have not received much attention (Deshpande et al., 2004; Sen & Bhatia, 2004). However, the momentum shifted when the question of economic well-being, unremunerative price of crops, indebtedness and farmer suicides were taken into consideration from early 2000. The growing agrarian distress became a central issue of the agrarian debate in India. Regarding the estimation of farm income, scholars have opined divergent views as different methodologies and data sources were employed. Some scholars have used CACP data, while others have drawn their arguments based on Comprehensive Scheme for Studying Cost of Cultivation (CCS) or GDP-based farm income data. Chand et al. (2015) have questioned the validity of the CCS data as it does not include the 'entire agricultural sector of the country' and ignores several minor crops like horticulture. However, farm income calculated from Situation Assessment Survey (SAS) data provides the actual annual income of farmer households. Analysis from the Situation Assessment Survey (SAS) data between 2002-03 and 2012-13 shows that most of the states have experienced low income from cultivation. Narayanamoorthy (2017) noted that even states like West Bengal, Bihar, Jharkhand, and J&K have observed negative growth of income from cultivation during the same period.

Farm households have different sources of income such as income from cultivation, income from livestock, income from wages and salary, and income from non-farm business. Farmers' income calculated between 2002-03 and 2012-13 based on SAS data conducted by the NSSO reveals the fact that income has improved considerably from Rs. 25,380 in 2002-03 to Rs. 77,112 in 2012. But the growth of income largely depends on the size of the agricultural land holding. It has increased marginally with significant variation from marginal to large-scale farmers having different access to land. Thus, compared to small and marginal farmers, the large landholders took less time to double their farm income (Satyasi & Bharati, 2016). A study conducted at the village level in three districts of West Bengal finds that income from crop production is very low. In potato farming, the presence of the middle man is making it a less profitable practice for the small and marginal farmers. Even in contract farming, the private companies do not like to engage small and marginal farmers due to the tiny size of their plots (Sarkar, 2017).

Different crops have different costs and profitability character. These are directly or indirectly related to the difference between the pre-reform period and post-reform period change in the cost of cultivation and farmers' income. From the analysis of wheat cultivation data between 1970-71 to 2004-05 of Haryana, Madhya Pradesh, Punjab, Rajasthan, and Uttar Pradesh, it is evident that the cost of wheat cultivation has increased three times faster during the reform period as compared to the pre-reform period. Fixed capital cost has remained stagnant, while variable costs have escalated, which has increased the total cost of cultivation. The cost for machine labour, chemical fertilizers and pesticides, seeds, irrigation, and interest has registered unprecedented growth since the 1990s because of the slashing of input subsidies and redundant state agencies (Raghavan, 2008). In a similar study, Dev and Rao (2010) showed that the growth rate in yield for paddy as well as wheat has declined after the 1980s. But after the 1990s cost of production has increased at a rate of nearly 1.5 percent per annum for rice and wheat. Due to fall in realized price of paddy in MSP, the profitability of paddy has been experiencing a declining trend while wheat farmers have experienced more profit in most of the years from 1981-82 to 2006-07.

Tripathi (2013) also opined that after 2004-05, Indian agriculture has experienced a sharp recovery from the slowdown that prevailed during the post-reform period, mainly due to non-price intervention by the government to make farming profitable enough. According to him, paddy prices received by the farmers were less than the MSP from 1998-99 to 2005-06, and after that, the price realized by farmers has remained more or less equal to the MSP of the paddy. However, in the case of West Bengal and Orissa, price realized by paddy farmers is still lower than the MSP even in the recovery phase. Tripathi (2013) finally concludes that wheat is more profitable than paddy cultivation. Apart from rice and wheat, the trend of profitability of five crops under rain-fed and irrigated conditions using the Commission for Agricultural Costs and Prices (CACP) data from 1970-71 to 2010-11 shows that *bajra* and maize cultivation is unprofitable, while cotton and groundnut cultivation are more profitable in the irrigated areas. Besides, most of the losses in cereal crops were incurred between 1995-96 and 2010-11, mainly due to the soaring cost of cultivation (Narayanamoorthy et al., 2014).

Declining farm income and productivity have serious implications on the growth of the Indian economy. The agrarian distress in recent years cannot be addressed unless the farmer income increases (Chand, 2016). From 1993-94 to 2004-05 farm income of cultivators dropped below 2 percent, but it increased by 7.3 percent in the next decade from 2004-05 to 2011-12. Thus, low growth in farm income is closely associated with agrarian distress and farmers' suicides. That is why the number of farmers' suicides has dropped after 2004 with higher agricultural growth (Chand et al., 2015). However, the aim of doubling farmers' income by 2022 has suffered largely from factors such as rising input costs, improper post-harvest management, absence of market infrastructure and irrelevant price realization, and non-remunerative prices in the market (Bihari et al., 2019).

Srivastava et al. (2017) have opined that between 2002-03 and 2007-08, agriculture was more profitable. Though the average annual inflation in A_1+FL cost (all expenses in cash and kind, plus value of family labour) was about 10 percent for the whole time period, it was comparatively lower during 2002-03 to 2007-08 and it witnessed a sharp increase after 2007-08 (about 13 percent). Farm mechanization also increases profitability and income. In the present situation, farm mechanization results in more output with less cultivation costs. However, farmers dependent on human labour suffer many shortfalls due to labour migration and labour engagement in the MGNREGA programme (Chandran, 2017). The question of profitability of farming is very closely interlinked with the question of household income. At the national level, crop profits have increased annually from Rs. 20,574 in 2002-03 to Rs. 31,015 in 2012-13. The growth rate was 3.7 percent per annum and it varies across states. Studies find that states with a fast rate of increase in household income have experienced profit with a higher rate and the correlation between household income and profit is very much positive (Vatta & Budhiraja, 2020).

From all these studies, it is clear that farmers, especially marginal and small farmers, all over India are not in a good condition. The profitability of agriculture has declined due to a rapid increase in input costs and due to slow growth in the price of crops. Freak weather, a highly volatile market, and lower realized price of crops have also aggravated the distress of the farmers. In such a situation, diversification to high value crops, farm mechanization, or increase in production is not the solution for sustainable agricultural growth.

Agricultural Distress and Farmer Suicides

Farmer suicide has been a common phenomenon in India during the post-reform period. In general, indebtedness, crop loss, low returns, and problems in agricultural marketing are the major causes of farmer suicides (Deshpande, 2002; Nirmala, 2003). According to the NCRB report, from 1995 to 2012, the farmer suicide rate in India was 13.9 percent among all suicide deaths (Mishra, 2014). Kennedy and King (2014) noted that farmer suicides will be higher in states where factors such as concentration of marginal farmers, proportion of cash crop production, and farmer indebtedness are much higher. Therefore, proper enactment of land reforms, stabilization in crop price, and controlling the indebtedness may be effective strategies against the farmer suicides. Sidhu et al. (2011) analyzed the pattern of farmer suicides in Punjab and identified certain reasons for the suicides. As per this study, 65 percent of total suicide victims belong to the age group of 20 to 40, as many as 79 percent of suicide victims are small and marginal farmers, and indebtedness is the major cause of committing suicide. In Maharashtra, empirical findings suggest that the majority of low and medium caste small farmers who committed suicide were adversely affected by the introduction of the neo-economic policies in the early 1990s (Mohanty, 2001).

Apart from crop failure, indebtedness, and rising costs of production, it was the aspired trend towards agro-based small-scale industrialization which trapped the small farmers to commit suicide (Mohanty, 2005). The state of Punjab is also in crisis, and many farmers have quit farming since 2000. Patil (2008) has shown that decline and uncertainty in earnings from agriculture, dependence on purchased inputs, and inadequacies in access to rural finance are the main causes of farmer distress. Though the interest rate of loans from non-institutional agencies is higher than institutional agencies, most of the farmers borrow money from non-institutional agencies due to lack of flexibility of the formal sector, delays in approval of loans, dominance of middlemen, and bribes on loan approval. The decline of long-term credit institutions has had a serious adverse effect on the small and marginal farmers (Patil, 2008). Singh (2012) recommends policy interventions for small farmers to cope with the agrarian crisis. It has been suggested that adequate investment, an agriculture relief fund, improvement in credit delivery, protection of farmers from distress sale, one-time settlement/compromise for non-wilful defaulters, and initiation of long-term agricultural credit policy are essential requirements for a revival (Patil, 2008).

For arresting farmer suicides and farmer distress, a rehabilitation package was announced on 1st July 2006 for 31 districts covering Andhra Pradesh, Maharashtra, Karnataka, and Kerala. Kalamkar and Shroff (2011), using secondary data and primary data collected from field surveys in Vidarbha, have studied the impact of the rehabilitation package. From the study, it is observed that farmers have benefitted from this package. Due to the proper implementation of packages, the number of livestock has increased significantly, and agricultural production has increased sufficiently.

Tenancy System

Agricultural tenancy is an important rural institution. Agricultural growth and farmer welfare have a direct relation with this institution. However, in the literature on agrarian relations, tenancy is one of the widely debated issues especially with reference to the agrarian economies of the Third World (Mohanakumar, 2014, p.1). Therefore, understanding the tenant-land owner relation is an important aspect of going deeper into agrarian relations.

Literature on the changing agricultural relationships in rural India is based on primary as well as secondary data. Using NSS data of various rounds related to ownership and operational holdings, it is found that it is not the pull factors of non-farm sector, but the push factor created by the non-remunerative agriculture that is responsible for withdrawing farmers from agriculture. Leasing-in land is responsible for the declining number of operational holdings in general, and an increase in the number of operational holdings among the farmers belonging to the size group of 0.4 to 1 hectare operational holding indicates growing leasing-in land by the marginal farmers. While medium farmers share a low proportion of their area under lease cultivation (Haque, 2021), farmer households with 2-10 hectares landownership are leasing out a major part of their total area. Among various kinds of tenancy, fixed money tenancy is increasing rapidly (Singh, 2006). However, "sharecropping was a predominant form of tenancy in the past" (Bansal & Grover, 2019). In his paper, Singh (2006) further opined that very high rents under fixed rent tenancy resulted in too small a net return, which is the main cause of farmer indebtedness and suicides in the year of crop loss. It is concluded that "contract farming has not been effective in all small and marginal farmers in its fold", and legalization of fixed rent tenancy may help to raise operational holding by leasing-in more land, as a further land reform has low potential.

Changes in the tenancy system lead to changes in cropping pattern, as observed by Mohanakumar (2014) in the context of Kerala, where farmers shifted from paddy to coconut farming. The study noted that land leasing on an annual contract is a common phenomenon in spite of legal prohibition. Mohanakumar (2014) also opined that lease cultivation is not always capitalist farming, rather it is a 'forced' one as agricultural labourers leased-in land to compensate for the loss in wage income resulting from the declining area under labour intensive paddy cultivation.

A review of tenancy relations will remain incomplete if tenancy relations in West Bengal are not discussed, as this state has a long history of struggle of tenant cultivators against landowners and landlords, and had experienced a successful tenancy reform programme which started in 1977, popularly known as *Operation Barga*. There are many studies on this reform and its impact on agricultural development in West Bengal. A comprehensive book on the topic of tenancy relations and agricultural development in West Bengal has been written by Bhaumik (1993). In the first chapter, the debate on the efficiency of share tenancy has been discussed. In the second and third chapters, historical processes through which the tenancy system evolved have been portrayed with the help of secondary information and NSS data of various rounds. In the subsequent chapters, excluding the conclusion, based on field data Bhaumik (1993) has discussed the participants in the land-lease market and their socio-economic status, different types of tenancy contracts and their terms and conditions, input use, cropping pattern, productivity, and returns from recorded and unrecorded tenancy.

In West Bengal, land reforms gave access to millions of acres of vested land to millions of poor, landless peasants and sharecroppers. In consequence, after the 1980s, the redistribution of land among the poor and landless households proved to be more productive than large-scale farming. This outcome shows that about 25-27 percent of the arable land was given out through land reform programmes. The redistribution of land, tenurial and institutional reform, through non-farm inputs availability and services helps the poor to increase their capacity for productivity and reduces poverty (Bandopadhyay, 2003).

In West Bengal, sharecropping tenancy is declining, while fixed rent tenancy is gaining importance due to willingness of both parties towards this tenancy (Bhaumik, 1993). Caste and level of education are also significant factors in the tenancy system, where a higher percentage of lessors belong to the upper caste and higher land-owning categories. Registered tenants enjoy a better position in setting down terms and conditions of tenancy, whereas unregistered weaker groups cannot bargain with the landowners (Bhaumik, 1993). *Operation Barga* has provided rights to the tenants against eviction, and better returns compared to the unrecorded tenants in spite of more or less the same productivity; thus, though tenancy reform has a role in agricultural development, it is not the sole cause.

Government Support Scheme

In India, agriculture has made a significant contribution to the country's national income, internal and external trade and commerce, food supply, and to industrial development. The government has taken many initiatives for the development of agriculture and for the betterment of the farmers' situation in India since independence.

In spite of technological and economic development, climatic variability and price fluctuation of crops make the farmers unstable in many parts of the country. To safeguard the farmers' interests, crop insurance policy is one of the most viable instruments to protect them from uncertainty and agricultural variability. The first pilot scheme, namely Comprehensive Crop Insurance Scheme (CCIS), was launched in 1985-86. In 1999-2000, it was replaced by the National Agricultural Insurance Scheme (NAIS). However, in India crop insurance covers a very small sown area and in terms of state-wise claims, it has shown a skewed pattern and wide regional disparity in claim ratio (Sinha, 2004).

However, due to delay in risk settlement, poor risk classification and 'moral hazard', the government has made certain changes in the National Agricultural Insurance Scheme in 2010. Other instruments include minimum support prices (MSP) and calamity relief fund.

In spite of growth of agricultural exports and procurement of paddy and wheat, farmers' income and profitability are still not conducive to minimizing the agrarian distress in the country. The Commission for Agricultural Costs and Prices (CACP) has recommended MSPs for 22 agricultural crops. The primary aim of introducing the MSP was to provide some sort of insurance to the farmers against the fall in market price (Chauhan, 2021). But evidences suggest that in comparison to deficit states, MSP was more effective for the surplus producing states such as Punjab, Haryana and Madhya Pradesh (Ali et al., 2012). Except for paddy and wheat, procurement through MSP is negligible for other crops such as millet, pulses, and oil seeds (Reddy, 2021). Researchers also find that low level of awareness and absence of a procuring agency (PA), are the major issues preventing the MSP from reaching its goal (Bhue & Kikon, 2020). In case of West Bengal, the MSP has a long way to go as a significant portion of farmers retail their crops below the MSP (Das, 2020). The Calamity Relief Fund (CRF) was also formed as per the recommendations of the Ninth Finance Commission to safeguard agriculture from different calamities.

In 1998, the Government of India along with the RBI, introduced Kisan Credit Card (KCC), a flagship programme to channelize agricultural credit flow. KCC provides short-term credit facility among the farmers. The main purpose was to meet their production and consumption needs. KCC also strengthened financial inclusion in the agricultural sector in India (Kaur & Dhaliwal, 2018). Findings based on eastern India suggest that KCC has helped to increase farmers' income, specifically among the small and marginal farmers. It has also reduced the dependency on moneylenders for borrowing of loans (Kumar et al., 2022). In spite of the progress and benefits, KCC has faced many operational challenges such as different eligibility criteria adopted by the banks, loans issued to the farmers only once, high stamp duty on higher loans, difficulty in crop-wise loan issued, farmers' awareness, etc. (Mehta et al., 2016).

The Book

The book is based on a micro-level research targeted to explore the emerging complex dynamics of agrarian crisis in India. To understand the whole gamut of the crisis, we have tried to focus on both farming as well as the farmer, because both are inseparable components of the agricultural economy, and we discuss various issues related to agriculture such as inputs, profit, tenancy, technology, lifestyle change, etc. Our target is to unearth the factors which are deepening the agrarian crisis in India. Besides analyzing many issues scientifically with appropriate methods, we have brought in subjectivity in understanding the crisis, especially from the farmer's perspective. We have not treated farmers as a single category and a uniform group of people, and instead of putting all farmers in a single box, we have tried to understand how the crisis is not the same for all categories of farmers, starting from big to small and marginal. In India, agriculture is increasingly not being practiced by the owner-farmers, leading to a huge proportion of short-term tenancy. To understand the crisis faced by these tenants of agricultural land, we have also considered those landless people who take other people's land on lease and cultivate crops.

By definition, agriculture includes plant cultivation and livestock rearing, while farming implies growing crops and keeping animals, and is a part of agriculture. In this study, both words (agriculture and farming) have been used to mean crop cultivation only, and other types of activities such as fishery, animal husbandry, etc., are not included. As paddy is the main crop in the study area, more importance has been given to paddy cultivation.

Central Ideas of the Book

Agricultural problems and their extent are not the same in all parts of India. Issues related to agriculture may differ from one state to another, and from one crop to another, in spite of having some common problems and issues. Different farmer categories may also be differently affected by the same problem. There may even be a considerable difference between the situation of a marginal farmer in Punjab or Maharashtra and a marginal farmer in West Bengal. Keeping in mind all these facts, the book addresses the central research question: How do the different issues such as globalization, technology use, tenancy system, change in labour and input conditions, change of rural culture especially of the farmer family, and government intervention affect the agricultural economy and deepen the agrarian crisis? To fulfil that central research question, the following issues are focused and are analyzed accordingly: a. The history of agricultural growth and farmers' situation in West Bengal as well as in erstwhile Bardhaman District; b. The problems related to major inputs with special reference to paddy cultivation; c. The profitability of cultivation in general and farmers' perceptions of profit in particular; d. Institutional support mechanisms and government initiatives taken for agricultural development; e. The negative impacts of technology use with special reference to tractor use; f. The gap between the household demand linked to socio-cultural change and farmers' financial status to fulfil those demands; g. The change in tenancy relations and their impact on agriculture.

The Research Process

The theoretical framework of the study combines different aspects (economic, political, environmental and socio-cultural) to highlight the crisis of farmers as well as of farming practice. Primary as well as secondary data have been used. Equal importance has been given to quantitative as well as qualitative data. Changes in the agricultural system and in farmer society have been shown in order to understand the growing intensity and complexity of the agrarian crisis in the selected area. In analyzing changes, the term 'in the past' has been used in many places to denote the situation before 1995. In digging the past, importance has been given to elderly farmers, who have long experience in the field and have maintained certain kinds of records such as cash books.

The Book is based on field research, and for empirical research we selected Galsi-I and Galsi-II Community Development (CD) Blocks of Purba Bardhaman district, which constitute part of the rice bowl of West Bengal. In selecting these blocks, we developed a composite score of all the CD Blocks of the Purba Bardhaman district on the basis of eight criteria: a. the percentage of irrigated area to total cultivable area, b. percentage of irrigated area under canal irrigation, c. percentage of deep tube well irrigated area, d. the yield of *aman* crop, e. crop concentration, f. the share of *bargadars*, g. percentage of agricultural labourers to total workers, and h. percentage of rural population served by banks. Galsi-I and Galsi-II CD Blocks ranked first and second respectively, which means that these two blocks have the highest level of all agricultural parameters and that is why we selected them. In these two blocks, paddy is the main crop and the yield is higher than the state average. Most of the agricultural land is irrigated. The infrastructure needed for agricultural development is also quite good compared to the other CD Blocks. Besides, '*Operation Barga*' has been successfully implemented in this area. Therefore, this area can be considered as a microcosm which represents the situation of farmers and the extent of agrarian crisis in West Bengal.

As there are 18 *Gram Panchayats* (GP), 18 villages have been selected covering one village from each GP using a purposive sampling method. Villages have been selected using secondary data sources and key informant interviews. In the selection process of sample villages, importance has also been given to certain parameters to minimize the risk of absolute subjectivity. These parameters are the distance from the railway station, distance from National Highway-2, availability of irrigation water, availability of bank service, and power supply for irrigation.

After selecting the villages, 20 farmer households from each village have been selected using a snowball sampling method. The unit of our study is 'farmer household', not just the 'farmer', only to capture the whole dynamics of the agrarian relations and the distress. Following the definition of NSSO 59th Round, by 'farmer household' we mean a "household having at least one member who possesses some land and performs agricultural activities in any part of that land during the 365 days preceding the date of survey".

In total, 360 farmer households have been surveyed using a semi-structured questionnaire for collecting primary data. The questionnaire had two broad parts: one for gathering information related to agriculture, and another for collecting socio-economic data on farmer households. Questions were set in such a way so that we could trace the change in agriculture and in farmer society over the last 20 years, because from that time the impact of liberalization started to be felt in the economy as well as in our society. The household survey was conducted in different phases in 2016-2018.

Besides the household survey, key informant surveys were conducted with the experienced and aged farmers to explore their perceptions on various issues, and with the Assistant Director of Agriculture (ADA) officials to gather information on how they calculate the cost and returns from crop cultivation, and to find out issues related to the ADA office. Focus group discussions were also carried out for collecting information on specific issues.

Secondary data have been collected from District Gazetteers of Bardhaman, District Statistical Handbook of Bardhaman, District Census Handbook, Primary Census Abstracts, NSSO reports on Situation Assessment Survey of Farmers, Economic Reviews, Agricultural data published by the Directorate of Economics and Statistics, and newspaper reports along with other academic sources.

For deeper understanding of the differential problems of the different classes of farmers in the agricultural practices, the farmer households have been classified into six categories based on possession of land. These are: a) landless lessees, who have no ownership of land, but lease-in land for a particular time period have been considered as landless lessee farmer households; b) landless sharecroppers, who have right to cultivate the owner's land and have to pay a share of crop as rent; c) marginal, with ownership holding of less than 1 hectare; d) small, with ownership holding of 1-2 hectares; e) medium, with ownership holding of 2-4 hectares; and f) big, with ownership holding of more than 4 hectares, to understand inter-class variation in perceptions among the farmers.

2 History of Agriculture in West Bengal

West Bengal has the sixth largest economy in terms of GDP in India. It has about 8 percent of the country's population, but produces more than 8 percent of the food grown in the country (Economic Review, 2017-18). The state is a leading producer of rice, potato, and vegetables, and also a significant producer of other crops and various fruits. Though rice and cereal production is in surplus, West Bengal is lagging behind in production of pulses and oilseeds (Economic Review, 2017-18). Nevertheless, agriculture is the "backbone of the economy of the state of West Bengal" (Ghosh, 2010, p.135). Agriculture contributed about 19 percent to the Gross State Domestic Product.

Like in India as a whole, the share of agriculture in the Net State Domestic Product is continuously declining in West Bengal, but the state continues to depend heavily on agriculture (Dasgupta & Bhaumik, 2014, p.107). As per the Economic Review (2017-18), there are about 72.81 lakh farmer households in the state. About 86 percent of the farmers in West Bengal are small and marginal (Gururaj et al., 2017).

Against this backdrop, this chapter attempts to examine agricultural growth focusing on area, yield, and production in West Bengal as well as in erstwhile Bardhaman district. Thereafter, the chapter tries to assess the farmers' situation from the historical past to the present.

This chapter is exclusively based on secondary data collected from different sources. The history of growth, decline, and revival of agriculture in West Bengal has been discussed based on literature written by Bose (1999), Dasgupta and Bhaumik (2014), Sarkar and Ghosh (2017), Ghosh (2008), among others. Besides, secondary data have been collected from the States Statistical Handbooks, Statistical Abstracts and Economic Reviews of different years. Reports on 'Situation Assessment of Survey' conducted by the National Sample Survey Organisation (NSSO) in the year 2002-03 and 2012-13 have been used to discuss the status of farmers in West Bengal. Newspaper reports have also been used to provide an outline on very recent agrarian problems and crises faced by farmers all over India as well as in Purba Bardhaman district of West Bengal.

History of Agricultural Growth in West Bengal

From the very beginning, West Bengal has been an agriculturally developed state. It has a history of long agrarian cycle of growth. However, this growth has been uneven. In the colonial era, from 1770 to 1860, almost all parts of rural Bengal experienced growth (Bose, 1999, p.42). However, between 1860 and 1920, agricultural production declined due to exhaustion of land of the moribund delta and malaria infection. During 1920-1946, agricultural output remained almost stagnant, at a rate of 0.3 percent, while the population grew at an annual rate of 0.8 percent in undivided Bengal (Bose, 1999, p.46). The situation in West Bengal was more pitiable. In Bardhaman division, the heart of rural West Bengal, agricultural output growth rate was -1.08 percent (Bose, 1999, p.46).

However, the situation started to improve after independence. West Bengal experienced a rapid growth in the agriculture sector. According to Boyce (1984), agricultural output grew at 2.74 percent during 1949-1980 in West Bengal. However, growth rate of per capita agricultural output was negative, which indicates that population growth was far ahead of agricultural growth. According to Gazdar and Sengupta (1999, p.61) agricultural output began to grow at a rapid rate in the 1980s and West Bengal achieved the fastest rates of growth in the output of food grains among all Indian states during that period. However, after the mid-1990s, agricultural growth rate started to decline all over India and this trend continued till 2004 (Alagh, 2013, p.7).

Therefore, based on the above discussion we may divide the post-independence history of agricultural development of West Bengal into ***four*** sub-periods.

a. Agricultural Development in the Pre-1980s Period

During this period, agricultural growth was stagnant in nature. Lack in the supply of water for irrigation, fertilizer, HYV seeds, electricity, credit, etc., were the major factors for low production. Nandy and Siddhanta (2000) estimated that during 1966-67 to 1976-77, rate of adoption of modern farming systems crossed 20 percent of the total rice area and annual growth rate was about 2.15 percent. During 1971-72 to 1982-83, the compound annual growth rate in production of food grains was -0.98 percent, while rate of growth in production of oilseeds and potato were 9.99 percent and 8.24 percent respectively (Nandy & Siddhanta, 2000). Growth rate of area, production, and yield of selected crops during 1970-71 to 1980-81, as calculated by Ghosh (2008), have been given in Table-2.1.

Table-2.1
Exponential Growth Rate of Area, Production and Yield of Selected Crops in West Bengal during 1970-71 to 1980-81

Crops	Area (%)	Production (%)	Yield (%)
Aus	-2.63	-4.37	-1.74
Aman	0.07	0.52	0.45
Boro	3.18	2.53	-0.65
Total rice	-0.2	0.51	-0.72
Total cereals	-0.31	-3.24	-2.92
Total pulses	-3.11	-3.16	-0.05
Total food grains	-0.51	0.35	0.86
Mustard	3.73	9.76	6.03
Total oilseeds	6.85	10.72	3.87
Potato	5.74	8.84	3.1

Source: Ghosh, 2008.

From Table-2.1, it is observed that growth rates of area and production for the *aus* and pulses were negative. Though *boro* rice output grew at the rate of 2.53 percent, growth rate of yield was negative. In spite of the Green Revolution in India during the mid-1960s, West Bengal failed to achieve considerable agricultural growth at that time mainly because of social and political institutions. During this period, there was a rise in the eviction of tenants and the resumption of land for self-cultivation by landowners (Gazdar & Sengupta, 1999, p.72). Enactment of the 'Bargadar Act of 1950', 'The West Bengal Estates Acquisition Act' of 1953, and 'West Bengal Land Reform Act of 1955' and their amendments created fear among landowners and eviction of *bargadars* started (Bhaumik, 1993, p.42-44). According to the report of the Land Revenue Commission, Bengal (1940) about 21 percent of the agricultural land was under *barga* cultivation during 1938 (Bhaumik, 1993). Therefore, it can be said that eviction of *bargadars* caused instability in agrarian society, which in turn resulted in sluggish and fluctuating growth during the mid-1960s and thereafter.

b. Agricultural Development during 1980 to mid-1990s

A tremendous change took place during this period. Crop production, as well as productivity of all major crops increased. Growth rates of area, production, and yield of some selected crops during this period have been presented in Table-2.2.

Table-2.2
Exponential Growth Rate of Area, Production and Yield of Some Selected Crops in West Bengal during 1981-82 to 1991-92

Crops	Area (%)	Production (%)	Yield (%)
Aus	-1.57	4.47	6.05
Aman	0.51	4.45	3.94
Boro	9.65	11.07	1.42
Total rice	1.23	5.31	4.07
Total cereals	0.97	7.18	6.21

Total pulses	-6.04	-4.31	1.8
Total food grains	0.57	4.52	3.93
Mustard	9.81	12.18	2.37
Total oilseeds	5.19	9.79	4.6
Potato	5.02	7.37	2.35

Source: Ghosh, 2008.

From Table-2.2, it is observed that growth rate of area, production, and yield of different crops during this period were higher than the previous period. However, growth of area under *aus* and pulses were negative. Yield of crops increased rapidly during this phase, which further resulted in production growth. Aside from pulses, other crops experienced positive growth rate in production. Production growth for *boro* paddy, mustard and potato was mainly an outcome of area growth than an increase in yield.

Dasgupta and Bhaumik (2014) also calculated that during 1980-81 to 1994-95, growth rate of area and production of total cereals were 1.05 percent and 4.80 percent respectively. Overall agricultural growth during this period was 5.39 percent per annum. Gazdar and Sengupta (1999) estimated that growth rate in food grains output during 1983-1992 was 4.29 percent per year.

However there is a debate regarding the impacts of various factors behind such growth. Banerjee et al., (2002) opined that *Operation Barga* "had a positive impact on agricultural growth" in West Bengal, and could explain around 28 percent of the productivity growth during this period. Chattopadhyay (2005) opined adoption of "new seed-fertilizer technology" by the marginal farmers "with the expansion of irrigation facilities" increased productivity, which in turn resulted in agricultural production growth. Saha and Swaminathan (1994) opined that along with the adoption of 'new technology' and implementation of *Operation Barga*, panchayati raj also played a major role in the improvement of agriculture during this period. To them:

"Panchayats are active in different production-related

activities. Panchayats are involved in water management; rural workers were organised to sink shallow tubewells to build contour bunds and to do irrigation-related earthwork of different kinds. Panchayats are reported to be involved in ensuring that cultivators receive electricity for agriculture, and in the allocation of rural credit. Panchayats are expected to distribute 'mini-kits' or input packages to small cultivators." (p. A10)

c. Agricultural Development during mid-1990s to 2004-05

Liberalization and globalization took place in this period and agriculture faced a declining growth trend again. Chattopadhyay (2005) opined that growth rate of rural employment declined, and income inequality increased during this period. Growth rates of some selected crops in terms of area, production, and yield have been presented in Table-2.3.

Table-2.3
Exponential Growth Rate of Area, Production and Yield of Some Selected Crops in West Bengal during mid-1990s to 2004-05

Crops	Area (%)	Production (%)	Yield (%)
Aus	-4.25	-2.38	1.89
Aman	-0.58	1.09	0.43
Boro	3.46	3.55	0.08
Total rice	-0.16	1.77	1.61
Total cereals	0.31	2.15	1.84
Total pulses	-0.63	0.35	1.07
Total food grains	0.27	1.94	1.66
Mustard	-0.33	0.86	1.19
Total oilseeds	0.32	0.21	-0.11
Potato	3.36	3.35	-0.01

Source: Ghosh, 2008.

From Table-2.3, it is observed that food grain production has declined during this period as compared to the previous period. Not only food grains, other types of crops also witnessed a slowdown in growth rate. It is also very interesting to note that yield of oilseed and potato experienced negative growth rate. However, growth rate of food grains production in India was lower than the population growth rate during 1999-2000 to 2004-05 (Sarkar & Ghosh, 2017, p.486). Against this backdrop, the performance of West Bengal was quite satisfactory, especially in food grain production.

Economic reforms in the 1990s resulted in major reduction in food and fertilizer subsidies, and imports of cheaper and subsidized agricultural commodities (Mukherjee & Chattopadhyay, 2017, p.501). Though developing countries like India had the opportunity to export agricultural commodities, they failed to do this because of the Special Safeguards under the WTO Agreement on Agriculture, and the high subsidies provided by the developed countries to their agricultural products (Bhalla, 2004).

Due to policy transformations as well as higher demand of high value food commodities, crop diversification started to gain importance in India as well as in the state of West Bengal. Dasgupta (2014, p.112) estimated that the area under vegetables and fruits grew at the rate of 1.70 percent and 5 percent respectively during 1997-98 to 2003-04, while production of vegetables and fruits grew at the rate of 2.46 percent and 5.75 percent respectively during the same period. However, Mukherjee and Chattopadhyay (2017, p.513-514) stated that as the canal irrigation system works on the basis of season and a time-bound schedule, it exerts a negative effect on vegetable production.

d. Agricultural Development after 2004-05

During this period, the area under lease cultivation started to increase after a declining growth trend observed over the past two decades. Besides, after 2004-05, public investment in agriculture started to increase and the National Horticulture Mission was launched in India. Favourable policy initiatives towards agricultural development resulted in increased yields (Deokar & Shetty, 2014). Exponential growth rate of area, production, and yield of different crops in West

Bengal during 2005-06 to 2014-15 have been calculated (Table-2.4).

Table-2.4
Exponential Growth Rate of Area, Production and Yield of Selected Crops in West Bengal during 2005-06 to 2014-15

Crops	Exponential growth rate (%)		
	Area (%)	Production (%)	Yield (%)
Rice	-0.38	0.319	0.706
Aus	-2.03	-0.901	1.154
Aman	-0.1	0.645	0.747
Boro	-0.912	-0.271	0.648
Total cereals	-0.322	0.482	0.806
Total pulses	0.652	2.192	1.669
Total food grains	-0.287	0.501	0.803
Rapeseed and Mustard	0.465	1.931	1.465
Sesame	1.187	1.701	0.379
Total oilseeds	0.587	1.59	1.298
Potato	0.387	4.027	3.623
Fruits	1.403	1.515	0.111
Vegetables	0.51	1.128	0.625

Source: Compiled from the data published by Directorate of Agriculture, Government of West Bengal.

From Table-2.4, it is observed that area under paddy cultivation has decreased, whereas production and yield have increased, but at a very insignificant rate. Moreover, area and production under *boro* paddy have witnessed a negative growth, and the growth rate of the yield is less than the growth rate of yield of *aman* paddy. However, potato, oilseeds, and pulses have registered relatively higher growth rates in terms of yield and production. Yield and production of potato have witnessed the highest growth among the major crops of West Bengal. Area-wise growth is highest for fruits and sesamum during this period.

The Rice Bowl of Bengal: Agriculture in Purba-Bardhaman District

Before Independence

The then district of Bardhaman (presently Purba Bardhaman) is one of the most agriculturally developed districts of West Bengal. Situated on a fertile alluvial plain, the district has always been agriculturally richer than the other parts of Bengal. It is known as the 'Granary of West Bengal'. Bardhaman's agricultural development dates back to the pre-Mughal era. During the Pal dynasty or during the period of Sasanka and the Sen dynasty, (1095-1206 AD) the district was completely dependent on agriculture (Samanta, 2001).

During the Mughal period, the district retained its fame in agriculture. However, *Jotedars*, *Zamindars*, and *Mahajans* used to control agriculture in the rural area, and rulers used to maintain relatively good relations with the farmers (Nandy, 2002). There were different classes of intermediaries. However, up to the period of Akbar, land revenue was relatively lower and thus the agricultural situation was better for the agrarian community. Though Mughal revenue administration was indirect and dependent on various intermediaries (McLane, 1993), Akbar planned to collect revenue directly from the farmers (Samanta, 2001). From the period of Jahangir, agricultural growth retarded, and the situation of the farmers started to deteriorate.

Some incidents like the *bargi* attack during 1740s, great famine of 1770s and malaria epidemics during the British period affected the agricultural development of Bardhaman. "In the decades of 1750s and 60s, Bardhaman gradually started to regain its lost agricultural prosperity and peasantry began to be rehabilitated" (Samanta, 2001, p.46). The farmers of the district suffered a lot again during the famine of 1770.

Different tenancy systems of the British rulers also exerted a negative effect on agricultural development in this district. However, due to effective measures taken by Bardhaman Raj, agricultural development in this district remained better compared to other places in Bengal (Samanta, 2001). Even after the implementation of permanent settlement, farmers of this district suffered least because of limited hierarchical layers of intermediaries. Throughout the 19th century, Burdwan (Bardhaman) district experienced a rise in the production, prices, and exports of food grains.

Bardhaman had an indigenous irrigation system which was badly affected due to the negligence of the British rulers (Samanta, 2001). According to Samanta (2001), laxity of the British government destroyed the overflow irrigation system of the district. In her words:

> "Due to the negligence of British rulers of old system of irrigation and embankment, the accumulation of silt completely blocked up the mouth of old channels of the Damodar. They did not take any action to clear up the passage by excavating the mouth of the old Damodar. As a result, a series of floods started to hit the district. Another important step of the British was to stop the breaches in the so-called *zamindari* banks to arrest flood. English rulers missed the point that many of these breaches actually served as safety valves conducting the accumulated pressure of a rising flood into the overflow channels." (p.48)

However, the British government also took some positive steps for the development of agriculture. To restrict the power of *Zamindars* and to save farmers from oppression, the government was compelled to formulate the Rent Act of 1859. Various tenancy legislations from that time provided some relief to the farmers. Between 1927 and 1931, the government also recorded the ownership of land. Due to all these steps, agricultural economy of this district started to revive from the beginning of the 19th century (Samanta, 2001). However, between 1920 and 1946, agricultural output growth was negative all over Bardhaman division (Bose, 1999). Factors such as spread of malaria epidemics, the great famine of 1943, rapid population growth, and exhaustion of the moribund delta land were responsible for the negative growth during this period.

Paddy and sugarcane were the two major crops of Bardhaman in the past. High-lying land around the village was used to produce a variety of crops, mainly to produce vegetables (Peterson, 1910, p. 90). Low-lying areas with clayey soil were used to produce paddy. *Diara* lands on the river banks were used for winter and spring crops such as pulses, oilseeds, wheat, vegetables, etc. Among the three types of paddy, *aman* paddy was largely cultivated. *Boro* paddy was produced on the sides of rivers, canals or lakes. High land was used to produce *aus* paddy. Peterson (1910) mentioned that potato was a profitable crop then, and that farmers of the neighbouring area of Chakdighi, Shaktigarh, and Satgachia achieved success in potato production. Agricultural implements and methods were very ordinary. Manure was widely used. Cow dung and cow dung ashes were mainly used as manure. Besides, mustard cake was used. Though there was no system of regular crop rotation, it was in use.

After Independence

Chakrabarti (1995) calculated that decadal growth rate of the area under crops during 1950-51 to 1960-61 was only about 4 percent, but the rate of growth of production during the same time period was 41 percent. In the 1960s the Intensive Agricultural District Programme (IADP) was introduced, and Bardhaman was selected as one of the IADP districts (Samanta, 2001). Implementation of this programme helped in agricultural development of Bardhaman district. During 1965 to 1980, exponential growth rate of agriculture was 3.85 percent in this district, whereas West Bengal experienced about 2.74 percent growth in agricultural output. Large-scale agrarian reform all over West Bengal helped in further growth of production in this district during the 1980s. Various favourable steps, such as expansion of irrigated area, co-operative movements, distribution of ceiling surplus land among the poor, tenancy reforms, etc., exerted a positive impact on agricultural development in the district of Bardhaman.

However, the negative impacts of the Green Revolution and economic liberalization are still visible in the present agricultural growth of Bardhaman district. This study has thus tried to analyze the trend of agricultural development in the district using data on area, production, and yield of some selected crops during different years (Table-2.5).

Table-2.5 clearly shows that total paddy production has experienced positive growth in terms of area, production, and productivity. However, there is a considerable difference in different kinds of paddy cultivation. After 1996-97, in spite of yield growth, *aus* paddy production has declined, caused by decrease in area under *aus* paddy. Area under *aman* paddy registered a declining trend after 1980-81, but production has increased due to yield growth. In case of *boro* paddy, the area has increased over time, but a decline in yield after 1996-97 resulted in a downturn in production. Wheat production has decreased after 1980-81. Oilseed production has slightly increased in 2005-06 compared to 1996-97, while potato production has slightly decreased. Observing the trends of growth in area, production and yield, it can be said that *aman* and *boro* paddy, and wheat cultivation are losing their importance, whereas potato cultivation is increasing in the state.

Exponential growth rates in terms of area, production, and yield of some selected crops produced in Bardhaman for the period of 2000-01 to 2013-14 have also been calculated to understand the present agricultural scenario (Table-2.6). During this period, except *aman*, *boro*, and potato, the area under other crops has experienced negative growth. Paddy production registered positive growth only because of *aman* paddy production. Pulses and mustard production has also declined. Yield of all selected crops registered positive but marginal growth. Therefore, it can be said that recent agricultural growth in Bardhaman district is quite stagnant in nature.

Table-2.5

Area, Production and Yield of Some Selected Crops in Bardhaman, West Bengal

Crops	Area ('000 hectares)				Production ('000 tonnes)				Yield (Kg/hectares)			
	65-'66	80-'81	96-'97	2005-06	65-'66	80-'81	96-'97	2005-06	65-'66	80-'81	96-'97	2005-06
Rice	456.4	493	626.2	639	657.3	708.8	1,741.40	1,968.50	1,514	1,438	2,781	3,081
Aus	23.6	23.8	32.4	14.6	27.5	43.5	79.9	44.6	1,059	1,824	2,470	3047
Aman	429.9	435.3	423.8	417.2	626.2	585	1,082.10	1365.5	1,382	1,344	2,553	3273
Boro	2.9	33.9	170	207.2	3.6	80.3	579.4	558.4	1,390	2,371	3,407	2695
Wheat	3.8	8.3	4.4	2.2	2.9	17.6	10.4	4.8	760	2,118	2,347	2199
Oil seeds	3.6	54.6	40.3	42.0	1.5	41.1	35.3	42.1	417	753	876	1001
Potato	16.2	26.1	41.1	43.4	192.2	548.7	1,243.20	921.20	12,779	20,993	30,233	21,249

Source: District Statistical Handbook, Bureau of Applied Economics and Statistics.

Table-2.6
Exponential Growth Rate in Area, Production and Yield of Some Selected Crops in the District of Bardhaman during 2000-01 to 2013-14

Crops	Exponential growth rate (%)		
	Area (%)	Production (%)	Yield (%)
Rice	-0.237	0.132	0.369
Aus	-3.522	-2.81	0.757
Aman	0.204	0.916	0.71
Boro	-0.9	-0.841	0.049
wheat	-5.182	-4.714	0.497
Total cereals	-0.26	0.112	0.373
Total pulses	-1.436	-0.949	0.484
Total food grains	-0.269	0.109	0.379
Rapeseed and Mustard	-3.75	-3.074	0.69
Total oilseeds	-1.622	-1.076	0.594
Potato	1.159	1.277	0.119

Source: Compiled from the data published by Bureau of Applied Economics and Statistics.

A comparative analysis of production of selected crops in different districts of West Bengal (Table-2.7) reveals that Bardhaman district ranks second in paddy production, ninth in oilseed production, and third in potato production.

Table-2.7
Production of Selected Crops in the Districts of West Bengal, 2015

District Name	Production in thousand tones			Share in state production (%)		
	Rice	Oilseed	Potato	Rice	Oilseed	Potato
Bardhaman	1896	42.4	1922.1	11.90	7.60	13.82
Bankura	984.2	28.8	1140.5	6.18	5.16	8.20
Howrah	320.4	16.8	293.3	2.01	3.01	2.11
North 24 Parganas	683.8	81	323.5	4.29	14.51	2.33
Nadia	780.9	139	218.4	4.90	24.91	1.57
Uttar Dinajpur	599.9	45.6	222.2	3.77	8.17	1.60

Malda	707.6	42.5	202.6	4.44	7.62	1.46
Darjeeling	68.5	0.3	122.3	0.43	0.05	0.88
Birbhum	1261.9	33	492.6	7.92	5.91	3.54
Purba Medinipur	1143.7	55	74.4	7.18	9.85	0.53
Hooghly	860	83.4	3506	5.40	14.94	25.21
South 24 Parganas	1018.7	17.2	39	6.40	3.08	0.28
Murshidabad	1190.5	124.6	392.5	7.47	22.33	2.82
Dakshin Dinajpur	520.8	26.6	123.3	3.27	4.77	0.89
Jalpaiguri	561.9	16.3	1199.9	3.53	2.92	8.63
Cooch Behar	720.1	8.7	1211.7	4.52	1.56	8.71
Purulia	599.4	3.1	10	3.76	0.56	0.07
Paschi Medinipur	2009.1	93.6	2413.5	12.61	16.77	17.35
West Bengal	15927.2	558.1	13908	100	100	100

Source: State Statistical Handbook, 2015.

From the above discussion, it is evident that although West Bengal, as well as Bardhaman district, has witnessed ups and downs in terms of agricultural growth, both are agriculturally developed. Let us now discuss the situation of farmers.

History of Agrarian Relations and Farmers' Distress

Farmers' distress is not a new thing. The agricultural relations and nature of agrarian situation have not remained the same during different periods of time. If we look into Indian history, it is observed that farmers have had to face deprivations in many ways. In this section, agricultural relations and farmers' conditions in India before independence have been briefly discussed. This part is mainly based on the book "*Land and All That, Land and Reforms in West Bengal: Perspective India*" written by Parimal Bandyopadhyay (2013).

Ancient India

In *Manusamhita*, it had been mentioned that at the time of establishment of Aryan civilization, farmers had to pay half of their produce to the king. However, later the amount of land revenue scaled down to one-sixth part or a twelfth part of produce mainly based on nature of soil. Besides, there were various *vali* or taxes, such as road cess, police cess, tax for settlement, etc. Though farmers were the majority, they occupied one of the least powerful positions in society (Bandyopadhyay, 2013). Singh (2013) opined that during the later-Vedic period, peasants were assigned a lower status, though it was not like that during the *Rigvedic* period. Farmers were dependent on rain, and in case of a failed rainy season they had to face various problems. However, normally it was believed that farmers were happy because of absence of intermediaries. There were four types of tenants: revenue paying tenants, rent free tenants, lease tenants, and sharecroppers. Around sixth century BC, the monasteries used to let out their land for *barga* cultivation (sharecropping). The King would help in creation and maintenance of sources of irrigation, and in turn would collect water tax, ranging from one-third to one-fourth of produce depending on nature of irrigation (Bandyopadhyay, 2013). Peasants and the King had a direct relationship during the Hindu period in the production system. Tenants had to pay revenue and taxes to the state. If the local administrator tried to collect any extra revenue or tax, he had to face strong action. Land revenue and taxes remained unchanged during the Vedic era. It was always advised to protect the farmers, as agriculture was the major source of revenue for the King. The feudal system started to develop during 300AD to 600 AD, and the practice of land grants in the age of the Guptas and Harsha led to the fragmentation of land and loss of right of peasantry to the land. Due to various taxes and obligations imposed by the intermediaries, free peasantry diminished and they faced misery.

Mediaeval Period (6th century to 16th century)

In the mediaeval period, the relation between landlords or king and peasants remained unchanged over a period of several centuries. The feudal system continued, along with the presence of sub-feudal landowners. As a result, peasants had to pay high rates of rent, and even had to pay an increased share of production. During the period of Mahmud and Balban in 13th century (1246 -1287), general land revenue was one-sixth of the total produce and in emergencies it was raised to one-fourth. However, Ala-ud-din Khalji abolished the age-old agrarian relations and raised the land revenue to fifty percent of gross produce as part of his measures to afford a strong army to combat Mongol attacks. As a result, peasants faced absolute poverty and misery (Bandyopadhyay, 2013, p.33). The Tughlaq rulers restored normal conditions again, where land revenue was reduced to one-tenth to one-twelfth of gross produce and revenue officials were instructed to consider the cultivators' capacity to pay. Besides, they improved irrigation facilities, which led to tremendous agricultural development. Thus, the Tughlaqs set the model and led the way in agricultural development (Bandyopadhyay, 2013, p.33).

Mughal Period

During this time the Bengal peasants had to pay one-third to half of the produce as revenue. Jagirdars used to collect surplus revenue and taxes to maintain the army. Mughal rulers had no direct contact with the people, which resulted in collection of various illegal taxes. Though cultivators were the main pillars of the economy and there was provision against ejection of the peasants, they had to face corporal punishment and were even compelled to sell off their wives or children if they failed to pay adequate revenue.

The feudal system continued in full form during the Mughal period. In Bengal, most of the feudal lords took the title *Raja* and sought to lead a luxurious life after paying tribute to the King. There was no fixed rate of revenue or taxes, and they would collect the maximum share of increased production. In this period feudal lords were named as *Jagirdars*. There were four kinds of *Jagirdars*: i) Nagdi jagir, ii) Tankha jagir, iii) Watan jagir, and iv) Paibaki. From the period of Emperor Jahangir, the condition of peasants started to deteriorate. He introduced a new system called *alatmagha* by which the collection of revenue directly from farmer stopped.

A class of local *Zamindars* emerged to collect the revenue. They started to exploit the farmers, which ultimately affected the agricultural production badly. Below the *Zamindars* there were *Ijaradars* and *Talukdars*. There were different types of peasants in the Mughal era. Among them four important categories are: i) *khudkhast* who had permanent possession of land, ii) *paikast* who were outsiders and had no hereditary right over land, iii) *bataidar* or *bhagchasi*, and iv) *kaljana*, *teyari*, or *balahar* who were landless peasants. Competition among the *Jagirdars* for retention of their authority led them to impose maximum tax and to repress the farmers.

In Bengal, Murshid Quli Khan's dynasty (1717-1727) was famous for devising methods for extracting maximum tax or levy or *abwab* from farmers. The next *Nawab* Suja-ud-din Khan and his successor Ali Vardi Khan imposed higher *abwab* without considering average production in cultivation. The revenue collecting agents also used to impose different kinds of taxes. Thus, it is evident that rent and taxes were too high for a peasant to pay. Another disastrous settlement introduced by Ali Khan fixed excessively high revenue demand based on this 'Hastabood'. Though a lot of agricultural development took place during the Mughal era, farmers also faced considerable coercion during this time period.

Colonial Period

Extension of irrigation, enactment of the Co-operative Act (1904 and 1912), and agricultural research were some of the major positive works of the British rulers for the development of agriculture. Despite these actions, the overall picture of agriculture remained bleak. In 1765, the East India Company got 'Diwani' of Bengal from the Mughal Emperor, and with this another chapter in the history of farmer oppression started. As a result, the great famine of 1770 occurred. Inspite of huge loss of population and considerable decrease of cultivation, the company collected land revenue in 1771 which was more than the revenue of 1768. This was just the beginning. In 1772, Governor Warren Hastings made a plan for a five-year settlement to ensure steady collection of revenue. Without assessing the productivity or the farmers' capacity, the company let the land out to the highest bidder for a five-year term. Later, in 1778, the annual settlement was put in place. These revenue settlements just aggravated the misery of the peasants. Lord Cornwallis made a ten-year settlement in 1790, which was converted into the permanent settlement in 1793. Under this settlement, revenue agents such as *Zamindars*, *Talukdars*, etc., were made proprietors of the soil and farmers became occupiers. This settlement also encouraged sub-infeudation, and as a result a class of intermediaries was created in between landlords and tenants. There were now four to five tiers of intermediaries between the *Zamindars* and tenants. In some places in Bardhaman, Birbhum, and 24-Paganas in West Bengal there were five tiers of intermediaries (Bandyopadhyay, 2013, p.60). Due to the presence of so many intermediaries, rent per acre was sky-high. Farmer eviction was a normal incident. Thus, peasants faced continuous and long-term misery. The administration was in favour of *Zamindars* and other intermediaries. When torture and exploitation reached beyond the limit, peasants either attempted to flee or organize themselves to revolt. All these factors resulted in economic depression for a long time from 1840 to 1860. Realizing the failure of the permanent settlement, the company enacted different laws to lessen farmers' destitution. The Bengal Rent Act, 1859, and Bengal Tenancy Act, 1885, gave relief to the farmers to some extent. However, due to several loopholes and higher rent, these laws also failed to stop the extortionate situation of the farmers. The history of farmers' revolutions from Sannyasi and Fakir Revolt to Tebhaga Movement clearly tells us that farmers were in a very bad situation under the British rule.

From the above discussion it is quite clear that there was a disproportionate relationship between agricultural development and farmers' welfare; rather, the farmers were in a distressed condition. From the Vedic period to the British period, affluence of the state and the rulers was an outcome of the hard work of the peasant class, but they were neglected. Generally, rulers took steps for agricultural development out of their own interests, not for the farmers' welfare.

Farmers' Situation in West Bengal and the Changes between 2003 and 2013

There are many studies on the effect of economic reform on agricultural development, but there was a lack of a pan-India study on the farmers' situation until 2003. National Sample Survey Organisation (NSSO), at the behest of the Union Ministry of Agriculture, Government of India, carried out the 'Situation Assessment Survey of Farmer household, 2003' (SAS 2003) in rural areas as a part of the NSS 59th round. The survey covered (SAS 2003) 51,770 farmer households from 6,638 villages all over India, of which 3,958 farmer households spread over 504 villages were selected from West Bengal. Five reports were published based on this survey. These were: i) Consumption Expenditure of Farmer Households, ii) Some Aspects of Farming, iii) Income, Expenditure and Productive Assets of Farmer Households, iv) Indebtedness of Farming Households, and v) Access to Modern Technology for Farming.

In 2013, NSSO in its 70th round of survey again conducted a Situation Assessment Survey of agricultural households (SAS, 2013). However, there are many differences in these two surveys. In the introductory part of these survey reports, all the differences have been highlighted. Besides, Sarkar (2017) has also made a comparison between these two surveys. In SAS 2003, possession of land was an essential criterion in defining farmer households, while a total income more than Rs. 3000 from agriculture over the last one year was an essential criterion in defining agricultural households. Therefore, these survey data are not strictly comparable.[3]

3. "Due to the change in coverage and difference in some important concepts and definitions followed in the two rounds, the estimates of SAS 2013 (NSS 70th Round) are not strictly comparable with those of SAS 2003 (NSS 59th Round). While making any comparison of estimates of these two rounds, sufficient care should be taken to account for these differences" (NSSO 70th Round).

Based on the reports published in 2003 and 2013, an attempt has been made to study the situation of farmers in West Bengal in relation to the all-India scenario. In this section, farmers' situation in West Bengal has been highlighted in brief mainly on the basis of income, expenditure, and indebtedness of different farmer categories.

Situation Assessment Survey of Farmer Households, 2003

According to SAS 2003 reports, about 57 percent rural households in West Bengal were farmer households, while 60 percent rural households were engaged in agricultural activity all over India. Average monthly income (excluding rent, dividends, etc.) per farmer household in West Bengal was Rs. 2,079, while at the all-India level it was Rs. 2,115. The highest average monthly income per farmer household was found in Jammu & Kashmir (Rs. 5,488) followed by Punjab. Thus, it can be said though average monthly income of the farmer households in West Bengal was close to the national average, it was less than half of the income generated by a farmer household in Jammu & Kashmir or in Punjab. Most of the income earned by the rural farmer households in West Bengal comes from wage/salary, followed by cultivation, while cultivation is the primary source of income for farmer households at the national level.

However, if we look into the different farmer households categories, it is observed that except for farmer households who have less than 0.40 hectare of land, cultivation is the most important source of income for the other categories of farmer households in West Bengal as well as in India (Table-2.8). Contribution of animal husbandry and non-farm business as a source of income showed a declining trend with increasing land possession. Thus, it is evident that farmers' income in West Bengal as well as in India is very low, and most of the farmer families depend on more than one source of income.

Table-2.8
Average Monthly Income (Rs.)
(excluding rent, interest, dividends, etc.)
per Farmer Household by Sources
during July'02 to June'03: West Bengal and India

Size class of land possessed (hectare)	Income from wages/ salary (Rs.)	Net receipt from cultivation (Rs.)	Net receipt from farming of animals (Rs.)	Net receipt from non-farm business (Rs.)	Total Income (Rs.)
		West Bengal			
< 0.01	996	0	45	393	1434
0.01 - 0.40	865	334	70	400	1669
0.41 - 1.00	943	1053	112	275	2383
1.01-2.0	686	2225	160	572	3643
2.01-4.0	978	4621	-33	427	5993
4.01 - 10.00	722	5263	-2465	344	3864
> 10. 00	NA	NA	NA	NA	NA
All classes	887	737	77	378	2079
		India			
< 0.01	1075	11	64	230	1380
0.01 - 0.40	973	296	94	270	1633
0.41 - 1.00	720	784	112	193	1809
1.01-2.0	635	1578	102	178	2493
2.01-4.0	637	2686	57	210	3590
4.01 - 10.00	486	4676	12	507	5681
> 10. 00	557	8321	113	676	9667
All classes	819	969	91	236	2115

Source: NSS Report No. 497: Income, Expenditure and Productive Assets of Farmer Households, 2003.

We may also turn our attention to the expenditure of the farmer households. From Table-2.9, it is observed that monthly per capita consumer expenditure (MPCE) of farmer households was lower than that of rural households in West Bengal as well as in India. The MPCE of farmer households in West Bengal and in India was about Rs. 508 and Rs. 503

respectively. Though MPCE of farmer households in West Bengal was close to the national average, MPCE for food items was higher and MPCE for non-food items was lower than the national average. Kerala was the state with farmer households having the highest MPCE. West Bengal ranked 19th all-India, in terms of MPCE of farmer households. All of these observations indicate that the financial condition of farmer households in West Bengal as well as all over India was very poor. Bhalla (2011) showed that there was minimal difference between MPCE of farmer households and households which were below poverty line (BPL). In case of our state, this difference was of just Rs. 147.

Table-2.9
Monthly per Capita Consumption Expenditure over Broad Groups of Items for Farmer Households: West Bengal and India

Item of consumption	Average MPCE of farmer households	
	West Bengal	India
Cereals	132.4	100.87
Gram, pulses and their products	12.37	16.57
Milk and milk products	18.83	48.71
Meat, fish and egg	46.86	15.7
Vegetables	38.85	30.6
Food total	311.32	278.74
Clothing	34.59	36.75
Education	19.08	16.83
Medical (Institutional)	8.27	10.71
Medical (Non-institutional)	23.61	23.69
Durable goods	15.54	18.57
Non-food total	196.42	224.09
Total consumer expenditure	507.74	502.83

Source: NSS Report No. 497: Income, Expenditure and Productive Assets of Farmer Households, 2003.

For the sake of precise investigation, ratios between average monthly income and average consumption expenditure among different farmer household categories have been calculated (Table-2.10). It is observed that the ratios for farmer households having less than 4 hectares of land are below unity in case of all-India level data, while in case of West Bengal it was only medium farmer households that had a ratio of more than unity. Ratio less than 1 indicates more expenditure than income.

Table-2.10
Ratio between Average Monthly Income and Average Monthly Consumption Expenditure by Farmer Household Categories: West Bengal and India

Size class of land possessed (hectare)	Ratio between average monthly income and average monthly consumption expenditure	
	West Bengal	India
< 0.01	0.62	0.60
0.01 - 0.40	0.72	0.68
0.41 - 1.00	0.8	0.68
1.01-2.0	0.94	0.79
2.01-4.0	1.26	0.97
4.01 - 10.00	0.74	1.23
> 10. 00	-	1.51
All classes	0.78	0.76

Source: NSS Report No. 497: Income, Expenditure and Productive Assets of Farmer Households, 2003.

In this context, higher rate of indebtedness among farmer households is a very common occurrence. Percentages of indebted farmer households were about 49 percent and 50 percent in India and West Bengal respectively. Andhra Pradesh with 82 percent of indebted farmer households ranks first in this aspect. Hence, indebtedness among the farmers in West Bengal is far less than the national average. In terms of indebtedness among farmer households, West Bengal ranked 13th in India. Moreover, the average outstanding loan amount per farmer household in our state is just Rs. 5,237, whereas the national level average is more than double (Rs. 12,585). Average outstanding loan was highest in the state of Punjab (Rs. 41,576 per farmer household). Thus, farmer households in West Bengal were not submerged in severe indebtedness. However, it may also be an indication of the failure of institutional sectors to provide loans to the marginal and small farmers.

Situation Assessment Survey of Agricultural Households, 2013

According to SAS 2013 reports, about 51.5 percent of the rural households in West Bengal were farmer households, while the percentage of rural households engaged in agriculture all over India was 51.9 percent. Andhra Pradesh has the highest percentage of agricultural households to rural households in India. About 91.5 percent agricultural households of West Bengal belong to marginal categories based on possession of land, whereas 69.4 percent agricultural households in India belong to these categories. Cultivation is the major source of income for 55.8 percent and 63.5 percent agricultural households in West Bengal and in India respectively. Wage or salaried employment is the major source of income of 26.8 percent agricultural households in West Bengal and 22 percent agricultural households in India.

The average monthly income (excluding rent, dividends, etc.) per farmer household in West Bengal was Rs. 3,980, while at the all-India level, it was Rs. 6,426. Most of the income (about 53%) earned by the rural agricultural households in West Bengal comes from wage/salary, followed by cultivation, while cultivation is the leading source of income for agricultural households at the national level. Cultivation is the second most important source of income for the agricultural households in West Bengal. It is observed that cultivation is the major source of income for the small and big agricultural households (Table-2.11).

Dependence on cultivation is lower in West Bengal than in India as a whole. However, monthly income of agricultural households in West Bengal is lower than the national average. Contrary to this, the percentage of agricultural households possessing BPL ration card is lower than the national average. In West Bengal, among agricultural households, about 31.7 percent have BPL ration cards and only 3.1 percent have *antodaya* ration cards, whereas at the all-India level about 36.4 percent agricultural households possess BPL ration cards and 4.9 percent possess *antodaya* ration cards.[4]

Table-2.11
Average Monthly Income (Rs.) per Agricultural Household by Sources during July'12 to June'13: West Bengal and India

Size class of land possessed (hectare)	Income from wages/ salary (Rs.)	Net receipt from cultivation (Rs.)	Net receipt from farming of animals (Rs.)	Net receipt from non-farm business (Rs.)	Average monthly income (Rs.)
			West Bengal		
< 0.01	1066	12	307	117	1502
0.01 - 0.40	2242	337	198	466	3264
0.41 - 1.00	1707	1514	301	864	4385
1.01-2.0	2227	3203	53	1131	6613
2.01-4.0	3464	3325	-95	1342	8036
4.01 - 10.00	21358	11165	4018	1557	38097
> 10. 00	0	5254	2222	5033	12510
All classes	2126.5	979	225	650	3980
			India		
< 0.01	2902	30	1181	447	4561
0.01 - 0.40	2386	687	621	459	4152
0.41 - 1.00	2011	2145	629	462	5247
1.01-2.0	1728	4209	818	593	7348
2.01-4.0	1657	7359	1161	554	10730
4.01 - 10.00	2031	15243	1501	861	19637
> 10. 00	1311	35685	2622	1770	41388
All classes	2071	3081	763	512	6426

Source: NSS Report No. 576: Income, Expenditure, Productive Assets and Indebtedness of Agricultural Households in India, 2012-13.

4. Ration card for poorest of the poor; this card holders get 35kg paddy and wheat per month at the rate of Rs. 2/kg for rice, and Rs. 3/kg for wheat.

From Table-2.12, it is observed that at the national level, ratios of average monthly income and expenditure per agricultural household for marginal agricultural household categories (possessing up to 1 hectare of land) are below unity but near about one. For rest of the categories, the ratio is more than one, i.e., income is more than expenditure. However, in case of West Bengal, the ratio between average monthly income and average monthly expenditure for all classes is far below unity, which implies that monthly expenditure is greater than monthly income. The pitiable situation of agricultural households belonging to the marginal categories in West Bengal is revealed if we look into the ratios for these size classes, as their earnings are much lower than the expenditure they incurred. In case of West Bengal, only medium and big agricultural households' income is greater than their monthly expenditure, but they comprise a very negligible share (only about 0.2 percent) of the total agricultural households in the state.

Table-2.12
Ratio between Average Monthly Income and Average Monthly Consumption Expenditure by Agricultural Household Categories: West Bengal and India

Size class of land possessed (hectare)	Ratio between average monthly income and consumption expenditure	
	West Bengal	India
< 0.01	0.29	0.89
0.01 - 0.40	0.62	0.77
0.41 - 1.00	0.68	0.87
1.01-2.0	0.81	1.14
2.01-4.0	0.90	1.38
4.01 - 10.00	2.62	1.94
> 10. 00	1.05	2.86
All classes	0.68	1.03

Source: NSS Report No. 576: Income, Expenditure, Productive Assets and Indebtedness of Agricultural Households in India, 2012-13.

As indebtedness is an indicator for assessing a household's financial condition, percentage distribution of loanee agricultural households and average amount of outstanding loan by size classes of land possessed have been presented in Table-2.13. In India, as well as in West Bengal, more than 51 percent agricultural households are indebted. At the national level, data reveals indebted households increase with the increasing size categories, but no such pattern is observed in case of West Bengal. There is a considerable difference in outstanding loan amount per agricultural household between West Bengal and India. In India, the average outstanding loan per agricultural household was about Rs. 47,000, whereas in West Bengal it is estimated as low as 2.5 times of the national average. However, there is a considerable variation in average outstanding loan among different size classes of land possessed at national level as well as state level.

Table-2.13
Distribution of Loanee Agricultural Households and Amount of Average Outstanding Loan by Size Classes of Land Possessed

Size class of land possessed (Hectare)	Loanee agricultural household (percent)		Average amount of outstanding loan (Rs.)	
	West Bengal	India	West Bengal	India
< 0.01	38.08	41.92	5700	31100
0.01 - 0.40	55.17	47.27	14600	23900
0.41 - 1.00	44.32	48.33	19700	35400
1.01-2.0	57.23	55.71	33000	54800
2.01-4.0	42.9	66.51	32900	94900
4.01 - 10.00	29.93	76.34	43500	182700
> 10. 00	100	78.74	276000	290300
All classes	51.53	51.94	17800	47000

Source: NSS Report No. 576: Income, Expenditure, Productive Assets and Indebtedness of Agricultural Households in India, 2012-13.

Therefore, from the brief analysis based on both Situation Assessment Surveys, it can be said that cultivation is not enough for sustaining livelihood. Financial condition of the farmers, specifically small and marginal farmers, in West Bengal as well as in India, is very poor.

Recent Agrarian Crisis in India and Farmers' Movement

Besides research papers and government reports, newspaper reports may also help us to understand this crisis related issue in a different way. Though these reports provide a piecemeal picture of the agrarian crisis in India, we can sense the recent agrarian crisis.

Moudgal (2016, December 12) in his report showed that farmers in the state of Karnataka were in a distressed situation during the period of 2014-17 mainly because of crop loss by drought. This crop loss resulted in indebtedness among the farmers and led to suicides of 1,476 farmers across Karnataka.

In a report published in *The Hindu* on 8th December, 2018, Jebaraj identified that indebtedness among the farmers, landlessness, and failure to generate jobs in other sectors are the major problems in the rural economy. Jebaraj also showed that fall of market prices of major crops in the years of record farm output, increase in input costs, and demonetization have together created the agrarian crisis, while lack of government initiatives have aggravated the crisis.

Rohit and Preetha (2019, June 16) in another report on sugarcane cultivation in Tamil Nadu highlighted that sugarcane farming has become unsustainable for the past four or five years due to scanty rainfall and declining groundwater. Sugar mill owners have also failed to make payments to the cane farmers because of sickness of the sugar industry. As a result of these two causes, indebtedness among farmer households is increasing, and many of them have been compelled to harvest other crops.

Reports on farmer suicides and unsustainable paddy cultivation in Bardhaman district were also published in national and local newspapers ("Unnatural death of farmer", 2014; "Suicide of two farmers", 2015; "Again paddy farmer's death", 2015; "Death of farmer at Bhatar", 2017; "Unnatural death of farmer again", 2017; "Farmer's death at Jamalpur", 2017; "Unnatural death of sharecropper", 2018). In most of these cases, causes for farmer distress were the low returns caused by the low market prices of agricultural products compared to the cost of production, natural calamities, and pest attacks. Among the causes of farmer suicide or death as reported by the family members of the deceased farmers, is the shock due to loss in *thika* cultivation coupled with informal loans.

Among the incidents of farmer protest in India, significant are the Kisan Long March in Maharashtra and the protests at Mandsaur in Madhya Pradesh. The Mandsaur protest started on 1st June, 2017 demanding fair prices for farm production, and compensation for crop loss caused by drought (*Financial Express*, 2017, June 7). Similarly, on 6th March, 2018, around 40-50 thousand farmers of Maharashtra started to walk towards Mumbai from Nasik. The major demands of this movement were implementation of a loan-waiver scheme for crop loss, pensions for the poor and aged farmers and agricultural labourers, reformation in public distribution system (PDS), proper prices (at least 1.5 times of the total cost of production), etc.

In Bardhaman, too, there are frequent farmer protests because of inadequate irrigation water supply, inefficiency of government procurement system, low realized prices of paddy and potato, etc. ("Farmers' agitation", 2017; "One year over, no payment" 2017; "Wastage of potato", 2017; "No payment" 2017; Chakraborty & Sarkar, 2018).

Conclusion

Agricultural development in India, in West Bengal, and in the district of Purba Bardhaman has experienced ups and down from the historical period onwards. Agricultural development was dependent on the policies implemented by the rulers for most of the time. In general, up to the British period, the rulers were keen for production growth from agriculture for their own interest, rather than farmers' welfare. During periods of crisis, the farmers' situation was miserable, but it cannot be said that for the rest of the time they were affluent enough or that they used to lead a happy and prestigious life. After independence, the Government of India took several measures to revive the agricultural economy since the first five year plan, but the policy reforms after the 1990s have once again made the farmers vulnerable to a higher degree. At present, marginality, very low income, indebtedness, etc., are the day-to-day experiences most farmers of India and of West Bengal are struggling with.

3

Major Inputs

Introduction

In the previous chapter we have discussed the agricultural growth and farmers' situation in India, in West Bengal, and in the district of Bardhaman. In this chapter we will discuss about the input use in agriculture. In economics, inputs refer to those items which are used in the production process to produce output or finished products.[5] In an agricultural system, inputs include seeds, fertilizer and pesticides, irrigation, labour, farm implements, knowledge, etc. All these inputs have great significance because use of these inputs influence production of a crop and partly determine profitability. In modern agriculture, availability and access of the agricultural inputs to the farmers are very important. Optimum use of these inputs is also essential to get the maximum advantages for a long time from cultivation.

The Green Revolution during the 1960s achieved huge success because of supply of subsidized high yielding varieties (HYV) of seeds, fertilizer and pesticides, and other inputs. Indian farmers are now accustomed to using these modern inputs. However, researchers have identified many constraints regarding these inputs. Nadkarni (1988) opined that higher rate of growth in input costs has made our agricultural production system unsustainable economically as well as ecologically. Nadkarni observed that the technological breakthrough in Indian agriculture has obviously increased the production per unit of land, but has failed to lower per

5. https://en.wikipedia.org

unit cost of production. Raghavan (2008) opined that agrarian crisis in India "has grown on a three pronged set of symptoms: rising input costs, dwindling produce price realization and the inability of farmers to abandon cultivation without alternative livelihood sources" (p.123).

Therefore, it is necessary to inspect the issues related to input use because this research topic is concentrated on the agrarian crisis. This chapter examines the trend and nature of changes in input use and the different issues related to these inputs such as seed, fertilizer and pesticides, irrigation, machine and human labour use, knowledge, and so on. This chapter is mainly based on primary data collected from field survey and interviews on input use. As paddy is the main crop in Galsi-I and Galsi-II CD Blocks, more emphasis has been given on the input use in paddy cultivation. Input use for crops like potato, mustard and sesamum has also been addressed, but in brief. Secondary data collected from different NSSO reports have been used to support the field-level observations.

Major Inputs of Agriculture

Land

Land is the first and foremost input in agriculture, and the size of landholdings is the most important and most debated issue in agricultural productivity related studies. As per Agriculture Census 2011, the average size of the holdings is 1.15 hectares, and about 44.58 percent land area is operated by marginal and small farmers. The major issue regarding the agricultural landholdings in India is its size and its relation to productivity. Generally, it is believed there is an inverse relation between the size of holdings and productivity. However, Sen (1964) opined this assertion is not well established yet. Chand, Lakshmiprasanna and Singh (2011), using secondary data collected from different sources like NSSO, Agricultural Census, and F.A.O., have shown the relation between farm size and yield based on the aggregate of all crops. Chand et al. (2011) compared the Indian scenario with China in terms of farm size, productivity, agricultural growth, and poverty. Classifying the landholdings as marginal, small, semi-medium, medium and large, they have shown the temporal changes in input use. They found input use such as access to irrigation, fertilizer use, adoption of high yielding varieties are highest in the case of the marginal farmers. It is opined that per capita output is lowest on smallholdings in spite of higher productivity because of relatively low per capita availability of land. At last, they suggested that employment generation for smallholders outside agriculture is needed to remove the poverty and to sustain agricultural growth because smallholders earn a meager amount of income from agriculture due to unfavourable man-land ratio.

Seeds

The seed is one of the basic inputs in agriculture. At present, demand for quality seed is increasing very rapidly. Reddy et al. (2005) have highlighted seed production in their study. According to them, the area under certified seed has increased from about 500 hectare in 1962-63 to about 6 lakh hectares at present, but the supply of quality seed is insufficient and so there is a scope of employment in the seed production programme. They suggested that the use of advanced technology, training of farmers, development of integrated seed industry, and initiation of *Seed Village Programme* are required for the improvement of seed quality and spread of quality seed.

The Government of India has introduced Seed Bill to ensure the supply of quality seed and to protect the farmer's right on seed production. However, Bhukta (2005) opined that this bill would harm the farmers. According to him, traditional seeds will disappear over time, and genetically modified seeds mainly supplied by multi-national companies will grasp the market, and as a result, farmers will be dependent on them for seed. Bhukta feared that complexity in the seed registration process would snatch away the farmers' traditional right to selling farm-saved seed.

Nevertheless, many recommendations have been made by Parliamentary Standing Committee on Agriculture in 2006 and based on these recommendations, Seed Bill, 2010 has been announced. Ravi (2010) has given thanks to PSCA for their recommendations helpful for protecting farmers' interest regarding seed use. However, like Bhukta, Ravi (2010) also opined that there is a lack of transparency in the registration process, have a scope of bio-piracy, and it will promote commercial monopoly.

Seed is a basic and the most important input in the agricultural system. Pure and good quality seeds are essential for production and for ensuring returns. Indian farmers used to produce crops using traditional seed varieties for a long time. The main problem with these traditional seeds was low yield. During the 1960s, HYV seeds become popular. "The increase in foodgrains production has been made possible mainly by the introduction of high-yielding varieties of crops which, under a given package of practices, are found to give a much higher yield than the traditional varieties" (Roy, 1971, p.A75). Use of HYV seeds spread rapidly within a short span of time and replaced traditional seed varieties. Traditional seed varieties are not generally used by farmers nowadays.

However, many scholars are suspicious about the gain from HYV seeds. Nadkarni (1988) opines that HYV seeds impose a threat to the genetic diversity of our plant life. It is also said that farmers have become totally dependent on the market for purchasing seeds because traditional seed varieties are no longer available to the farmers and they are now habituated to using HYV seeds (Kumar & Singh, 2016, p.7015). Moreover, farmers have to replace the seed varieties with newer ones every three to four years in order to keep the production intact. Other important concerns are the quality of the seed varieties and sufficient supply of seeds. National Seed Corporation, a public sector entity under the Ministry of Agriculture, was the main producer of HYV and hybrid seed varieties in India. The role of private sector in seed production was negligible for a long time. Later, signing of the World Trade Organization

(WTO) in 1995 paved the way for private seed companies to enter into the seed market. Several multinational seed companies are controlling the seed market in India at present. Apart from HYV seed varieties, genetically modified (GM) seed varieties have also made their way into Indian agriculture. However, many reports (Shiva, 2013; Tendulkar, 2019) suggest that GM seeds have also failed to increase production at the desired level. Besides, GM seed use has resulted in a monopoly market condition. On entry of multinational seed companies and use of GM seeds, Dr. Shiva (2013) in one of her reports opined that, "a renewable resource (seed) became a non-renewable, patented commodity". It is also argued that legislative measures taken by the government also favour these multinational seed companies. In another article Shiva (2005) claimed that, "the Seed Act 2004 has one and only one objective of stopping farmers from seed saving, seed exchange and seed reproduction".

However, in this research issues related to GM seeds have not been addressed. We have only asked questions to the farmers about what seed varieties they use; from where they get the required seeds; how they replace seed varieties; and what kind of problems they face related to quality seed. We were compelled to exclude GM seed or seed bill related issues from the questionnaire as farmers of the study area are ignorant about these issues, and GM seed varieties are not in use here.

Changing Pattern of Seed Varieties

Paddy Seed Varieties

During the 1980s and 1990s, farmers of the study area used to grow CR 1010, *Ratna*, *Pankaj*, and IR 8 varieties of paddy. These were HYV seeds of paddy at that time. However, traditional seed varieties such as *Shu Kalma*, *Dudh Kalma*, *Jiakathi*, *Jhinge Sal*, *Dohornagra*, *Khejurchori* etc. were also grown by the farmers. During 1991-92, about 59 percent of the area under paddy cultivation in West Bengal was under HYV seeds (Chand & Haque, 1998, p.A110). According to the farmers interviewed, these traditional paddy varieties were of good taste and required less fertilizer, but productivity was very low (only 6-8 sacks of 60kg/bigha). Bigha is a local unit of land measurement which is around one-third of an acre. Due to low yield rate, these traditional paddy varieties became obsolete.

At present the most used seed varieties of paddy in the study area are *Lal Swarna* or MTU 7029, MTU 1010 or *Cottondora Sannalu* and IET 4786 or *Satabdi*. All these seed varieties are HYV seeds. MTU 7029 is known as *Lal Swarna* to the farmers and is cultivated during *kharif* season. Earlier, IET 5656 seed variety was in use. That variety took 150-155 days for maturity while MTU 7029, the modified variety of IET 5656, takes only 140 days to mature. IET 5656 was known as *sada swarna* to the farmers. Yield rate of the *Swarna* varieties is about 55-60 quintals per hectare.[6] As MTU 7029 takes less time to mature, this variety is highly used at present in place of IET 5656. Other seed varieties grown in the study area during *kharif* season are Gotra Selection (GS), MTU 1010 and IET 4786. Yield rate of GS variety is high, but being a short duration variety there is a chance of seed germination if wet weather prevails during harvesting period, that is, during the first half of October. Besides, yield rate of this paddy is lower than *swarna* paddy. On the other hand, MTU 7029 is a medium or long duration seed variety. That is why *swarna* paddy is the most popular paddy variety to the farmers for *kharif* season. MTU 1010 and IET 4786 are also grown in part of the cultivated area during *kharif* season by some farmers who either plan to produce *rabi* crop in the same land after harvesting paddy, or plan to reproduce seed for the *boro* season.

During *boro* season, IR 36, MTU 1010 and IET 4786 seed varieties are largely grown by the farmers. All of these varieties are of short duration varieties, that is, their maturity period is less than 120 days. Average yield rate of IR 36 is 40-55 quintals per hectare. Average yield rate of MTU 1010 is 70 quintals per hectare, while average yield rate ranges from 35

6. http://www.rkmp.co.in

to 56 quintals per hectare for IET4786 variety.[7] IET 4786 variety of paddy is known as *minikit* paddy to the farmers as well as to the common people.

Majority of the farmer households cultivate *swarna* seed varieties only during *kharif* season. About 74.86 percent farmer households of Galsi-I CD Block and 77.77 percent farmer households of Galsi-II CD Block grow only *swarna* variety of paddy during *kharif* season. The rest of the farmer households used to grow more than one variety of paddy during this season. Paddy varieties grown during *kharif* season by the farmer households of the two CD Blocks are shown in Table-3.1.

Table-3.1
Distribution of Farmer Households by the Use of Paddy Varieties during Kharif Season

Paddy varieties	Farmer households (%)	
	Galsi-I	Galsi-II
Only Swarna	74.86	77.77
Swarna+IR 36	10.86	4.44
Swarna+Gobindobhog	4.57	3.89
Others	9.81	13.9
Total	100	100

Source: Field Survey, 2016-18.

It is observed that 5.14 percent farmer households of Galsi-I and 7.22 percent farmer households of Galsi-II produce *gobindobhog* paddy in part of their land. In this connection, it should be mentioned that *gobindobhog* is a special variety of paddy with a geographical indication (GI) tag of West Bengal and is cultivated mainly in Raina-I, Raina-II and Khandaghosh CD Blocks of Purba Bardhaman district (Roy, 2018). Since it is more profitable than common varieties of paddy like *swarna*, farmers of the study area are also becoming interested in its production. Zakir Hossain, a small farmer from Sankrai village of Galsi-II CD Block said that he cultivates *gobindobhog* paddy as it fetches more profit even

7. http://drdpat.bih.nic.in

though productivity is lower (9-10 bags per *bigha*) than common paddy.

Aman paddy is produced during *kharif* season by all the farmers, but *boro* paddy is not cultivated by all of them. About 77.77 percent farmer households of Galsi-I CD Block and about 80 percent farmer households of Galsi-II CD Block produce *boro* paddy. In Galsi-I CD Block, the most used seed variety of this season is IR 36 followed by MTU 1010; while for Galsi-II CD Block MTU 1010 is the most favoured seed variety of *boro* season (Table-3.2).

Table-3.2
Distribution of Farmer Households by the Use of Paddy Varieties during Boro Season

Paddy varieties	Farmer households (%)	
	Galsi-I	Galsi-II
IR 36	44.29	19.44
MTU 1010	22.86	37.5
IET 4786	9.29	10.42
IR 36+IET 4786 Mixed varieties	23.56	32.64
Total	100	100

Source: Field Survey, 2016-18.

During *boro* season, the most favoured paddy varieties are IR 36 and MTU 1010. Most of the farmer households of Galsi-I grow IR 36 paddy, while MTU 1010 is cultivated by most of the farmer households of Galsi-II. IET 4786 paddy cultivation is growing rapidly in the study area. The reason for this is that IET 4786 is a fine grained paddy, whereas IR 36 and MTU 1010 are coarse grained paddy. Being fine grained, IET 4786 paddy provides more returns to the farmers in spite of lower productivity than IR 36 and MTU 1010. In many villages, a few farmers are now cultivating Basmati varieties in place of high yielding varieties. Chand and Haque (1998) had observed such a trend in Punjab. This is because although Basmati varieties have low yield potentials, these are more remunerative due to fast rising demand for export, and they also require less water than HYV seeds (Chand & Haque, 1998, p.A110).

Potato and Mustard Seed Varieties

Apart from paddy, mustard and potato are two major crops of the study area. For potato, *Kufri Jyoti* and *Kufri Pukhraj* varieties are produced by the potato growers. Of these two varieties of potato, yield of *Kufri Pukhraj* is higher and it takes lesser time to mature than *Kufri Jyoti*, but *Kufri Jyoti* is more resistant to wet weather conditions. In general, *Kufri Pukhraj* variety of potato matures in 70-90 days and average yield is 130 quintals per acre, while *Kufri Jyoti* variety matures in about 90-110 days and yields about 120 quintals per acre.[8] *Kufri Jyoti* is an older variety of potato than *Kufri Pukhraj*. Generally farmers grow *Kufri Jyoti* potato for their own household consumption, and grow *Kufri Pukhraj* for selling in the market. Farmers prefer to produce *Kufri Pukhraj* for its higher productivity and/or for its shorter maturity period. Farmers who aim to produce *boro* paddy after harvesting potato, prefer to grow *Kufri Pukhraj* potato.

Farmers do not know the name of the variety of mustard they produce. According to them, two types of mustard are cultivated in Galsi-I and Galsi-II. One is *sada sorisha*, and the other is *kalo sorisha*. *Sada sorisha* is of better quality in terms of oil content than *kalo sorisha*, but some farmers prefer to produce the latter as it is less prone to pests and domestic animal attack, and requires less irrigation.

Source and Availability of Seed

Source of seed is an important aspect of seed use in agricultural crop production. According to the Situation Assessment Survey (SAS) 2003, about 47.8 percent farmers in West Bengal used purchased seeds, whereas about 48.1 percent farmers in India used purchased seeds. As for the usual way of procuring seeds, 62.22 percent of farmers were estimated to depend on purchased seeds, while the rest of the farmers used either exchanged seeds or farm saved seeds for paddy cultivation before 1995 in Galsi-I and Galsi-II CD Blocks. Therefore, it is observed that farmers of this area

8. https://www.ikisan.com

were far ahead in using purchased seeds as compared to farmers of the state, as well as of the all-India average. During our survey in 2016, about 88.6 percent of the farmers reported that they depend on purchased seeds for paddy cultivation. This change indicates that farmers have become more dependent on purchased seed varieties. Farmers responded that they prefer to use purchased seed for getting higher as well as assured production.

In the state of West Bengal, State Seed Corporation Limited (WBSSCL) organizes seed production through its registered growers, seed producers, NGOs, government farms and the State Agricultural University for *kharif*, *rabi* and *zaid* season every year to meet the demand of farmers as well as the government.[9] In 2014, West Bengal has also launched the "State Seed Mission" to meet the demand for seeds in the next five years ("Bengal Launches Seed Mission", 2014). In the district of Purba Bardhaman, there are many government seed farms, co-operatives which produce certified seeds and/ or sell seeds to the farmers. There are a government seed farm and a Krishi Vigyan Kendra (KVK) under the Central Government in Budbud of Galsi-I CD Block. Besides, there is a Crop Research and Seed Multiplication Farm (CRSMF) under the University of Burdwan, which is also engaged in producing foundation and certified seed varieties of different crops.

Though there are a number of public sources of seeds, farmers of the study area purchase seed from different sources, mostly from the markets. We have studied in detail about the sources of paddy seeds for the farmers, which have been presented in Table-3.3. It is observed that only 34.45 percent farmers purchase paddy seeds from public sources such as co-operatives, seed farms, Agricultural University, KVK, etc., while about 7.78 percent farmers purchase seed from Burdwan seed farm located near Kalna Gate as well as from CRSMF. Few farmers also buy paddy seeds from seed farms of Sonamukhi in Bankura district and from Bidhan

9. http://www.wbsscl.com

Chandra Krishi Viswavidyalaya of Nadia district. More than 50 percent farmers purchase seeds from the market. Among these farmers, about 36.11 percent farmers collect seeds from local markets. Galsi, Budbud and Ramgopalpur are three important local markets in the study area from where farmers purchase seeds.

Table-3.3
Distribution of Farmer Households by Source of Paddy Seeds

Usual source of paddy seeds	Farmer households (%)	
	Before 1995	At present
Co-operative	5.28	10.56
Budbud Seed Farm	1.95	9.44
Krishi Vigyan Kendra (KVK)	0.28	2.78
Seed farm of Burdwan and CRSMF	2.50	7.78
Burdwan market	11.94	13.06
Local market	29.17	36.11
From other districts	1.11	1.95
Aratdar/Mahajan	5.00	5.00
Other farmer/Own farm saved seed	37.78	11.39
Seed farms and local market/ own farm saved	0	1.94
No response	5.00	0
Total	**(100)**	**(100)**

Source: Field Survey, 2016-18.

The change in the distribution of percentage of farmer households by sources of paddy seed is also pronounced. Before 1995, about 37.78 percent of the farmer households were accustomed to exchanging seeds or using their own farm saved seeds for paddy cultivation. Only about 29.17 percent farmer households were dependent on local markets for paddy seeds. However, at present only 11.39 percent farmer households reported that they either use exchanged seeds or use their own farm saved seeds for paddy cultivation. Therefore, a trend of increasing dependency on the market for paddy seeds is observed. Besides, performance of public agencies as a source of seeds to the farmers has also improved. However, it is very interesting to note that the role of *aratdar/ mahajan* as a source of seeds to the farmers has remained unchanged.

A distinct spatial pattern is also evident from the distribution of farmer households by source of seeds (Table-3.3). As Budbud is near most of the villages of Galsi-I CD Block, farmers of these villages enjoy more advantages in purchasing paddy seeds from KVK and from Budbud seed farm. Farmers of Galsi-II CD Block are more dependent on Bardhaman seed farm and CRSMF among the public sources of paddy seeds. The farmers of Galsi-II CD Block are less dependent on the local market for paddy seeds than the farmers of Galsi-I CD Block, because they get paddy seeds at relatively low cost from Bardhaman market.

Farmers of these two blocks are fully dependent on private sources for seeds of other crops. Sometimes, the government provides free of cost seeds for crops like mustard, lentil, chick peas, etc., but supply is very limited and very few of the farmers get the scope to collect it. Many farmers complained that in most cases these seeds were provided to farmers who do not actually use them to grow crops, but rather use them for household consumption.

Most of the potato growers depend on either local markets or local traders for potato seeds. A few of the potato farmers collect potato seeds from farmers or relatives in neighbouring Bankura district. In many cases, potato growers of a village collectively decide from where they would buy potato seeds. Such collective decisions help them to reduce the carrying cost of potato seeds. Contract farming is also practiced by some of the potato growers in Taherpur village in which Pepsi Co. provides special potato seed varieties to the farmers, and provides assured prices for the produce.

Seed Replacement

The number of generations up to which the seed can be used from the previous crop is another important aspect, vital for the maintenance of crop productivity (Verma & Sidhu, 2009). Seed quality goes down with increased use due to various causes, and productivity of a crop also declines with deterioration of seed quality. This deterioration is slow in self-pollinated crops like paddy (Verma & Sidhu, 2009, p.326). Desired seed replacement rates are 25 percent for self pollinated crops, 35 percent for cross-pollinated crops and 100 percent for hybrids.[10] The seed renewal period as recommended by the National Commission on Agriculture (1976), is four years in paddy (Verma & Sidhu, 2009, p.323). Generally, seed replacement is measured using a ratio, called seed replacement ratio (SRR). SRR denotes how much of the cropped area is sown with certified seeds compared to farm saved seeds. In West Bengal, seed replacement ratios (SRR) were 35 percent for paddy, 50 percent for mustard, and 33.04 percent for potato during 2013-14.[11] All these figures are close to or above the target set for the 12th five year plan period.

In India, as per SAS 2003, about 30.33 percent farmer households replaced seeds every year; 31.57 percent farmer households replaced seeds every alternate year; 21.01 percent farmer households replaced seeds after three years, and 17.08 percent farmer households replaced seeds after four years or more. However, in West Bengal about 30.09 percent farmer households replaced seeds every year; 39.71 percent farmer households replaced seeds every alternate year; 17.02 percent farmer households replaced seeds after three years and only about 13.17 percent farmer households replaced seeds after four years or more. Therefore, it is observed that farmers of West Bengal were more advanced, as about 70 percent farmers replaced seeds within two years. Certainly, seed replacement has increased over time. Fearing production loss, nowadays the farmers in West Bengal normally replace seeds after using them for just one or two times.

10. https://seednet.gov.in
11. https://www.rkvy.nic.in

SRR has not been calculated in this study, but responses on the question about how many time farmers used farm saved paddy seeds have been presented in Table 3.4. As no significant differences in seed replacement practices among different farmer household categories of Galsi-I and Galsi-II CD Blocks are observed, only the combined result has been presented. It is observed that 15 percent farmer households use only certified seeds every year for paddy cultivation in the whole of their cropped area (Table-3.4). They do not use farm saved seeds or truthfully labelled (TL) seeds . The rest of the farmer households use certified seeds as well as farm saved seeds or TL seed[12].

Table-3.4
Size Category wise Distribution of Farmer Households by Use of Paddy Seeds

Farmer household category	Farmer household			
	Each and every year use certified seed only	Produce TL seeds and replace after using one time	Produce TL seeds and replace after using two times	Produce TL seeds and replace after using three times
Landless lessee	3 (20)	7 (46.67)	3 (20)	2 (13.33)
Landless share-cropper	2 (10)	7 (35)	8 (40)	3 (15)
Marginal	29 (15.42)	110 (58.51)	39 (20.74)	10 (5.32)
Small	8 (10.39)	47 (61.03)	20 (25.97)	2 (2.60)
Medium	5 (12.19)	25 (60.98)	9 (21.95)	2 (4.88)
Big	7 (36.84)	8 (42.11)	3 (15.79)	1 (5.26)
All	54 (15)	204 (56.67)	82 (22.78)	20 (5.56)

Source: Field Survey, 2016-2018; percentage is given in parantheses.

About 56.67 percent farmer households produce farm saved seeds and use them for one season either in the whole or most part of their land under paddy cultivation. They

12. Seeds produced by the farmers, for which seed certification is not required are called truthfully labelled seeds.

follow one of the three strategies for generating farm saved seeds: a) use certified seeds in part of the land every year, use the previous year's farm saved seeds in the rest of their land, and save paddy produced from certified seed as seed for using next year; b) produce paddy using certified seeds in a small part of the land during *aman/boro* season, and save the production as farm saved seeds for use in the coming season; c) use certified seeds in the total area under paddy cultivation and next year use only farm saved seeds generated from that certified seed variety.

Farmer households who use farm saved paddy for two and three times were estimated at about 22.78 percent and 5.56 percent respectively. However, foundation and certified seeds can be multiplied up to stage I and II.[13] Therefore, farmers who use TL seeds of 3rd cut may have low production because of deteriorating seed quality. Generally, farmers with poor financial conditions and/or having lack of knowledge about seed quality deterioration use TL seeds of 3rd cut.

A variation is observed among different farmer household categories in terms of seed use. It is the big farmers who rank first, followed by lessee farmers among farmers using certified seed only. About 36.84 percent of the big farmer households and 20 percent of the lessee farmer households practice 100 percent seed replacement in the study area. Since they are in a better financial condition, big farmers are able to purchase certified seeds for their total cultivated area under paddy. On the other hand, lessee farmers use certified seeds in order to fetch more profit through higher production. Small farmer households, followed by medium and marginal farmer households, rank first in use of first cut paddy seed. In second cut and third cut TL seed use, sharecropper households rank first. About 40 percent and 15 percent of the sharecropper households use second cut and third cut of a certified paddy seed variety.

For potato cultivation, most of the farmers use either

---

13. http://vikaspedia.in

certified seeds or first cut TL seeds. Many poor potato farmers also use second cut of TL seeds. However, for other crops like mustard or green vegetable cultivation, farmers use certified seeds only. Since seed requirement is very low for these types of crops, so even poor farmers can afford to buy the certified seeds.

Seed Rate

The seed rate or seeding rate is the number of seeds planted per unit area. It is expressed by number of germinating seeds and the weight of the seeds. Seed rate in paddy cultivation may vary from farmer to farmer based on seed varieties and quality of seed they use, methods of paddy cultivation they follow, nature of nursery bed, season of paddy cultivation, spacing and number of seedlings in plantation stage, and so on. Seed rate becomes low if a farmer cultivates small grained paddy. For example, *gobindobhog* paddy requires lower amount of seed per unit area as it is of small grained paddy. Seed rate for certified seed varieties of paddy are low compared to farm saved seeds because of good germination and healthy seedling generation from certified seeds. Besides, lower seed rate is required in *aman* paddy season than in *boro* paddy season because of higher rate of germination in the hot and humid summer season. There is an inverse relation between seed rate and spacing of seedlings transplantation, but a proportionate relationship exists between the number of seedlings per plantation and seed rate. More seed is required in the plantation method than in broadcasting or system of rice intensification (SRI) method. Seed requirement also varies with the nature of nursery bed. Nursery beds are of two types: dry and wet. If seeds are sown in a dry nursery bed, seedlings become stronger than seedlings of wet bed, and farmers thus require fewer number of seedlings per unit area for plantation. Dry nursery bed also requires less water.

However, in dry nursery bed, seedlings take a longer time to grow, grow more weeds, and require more number of labourers to transplant seedlings. Thus, farmers prefer wet seed bed to grow paddy seedlings. Such a choice certainly saves time but makes the input cost for seeds higher. In the study area, average seed rate is 8-10 kg per *bigha*, though variation in seed rate is observed among farmer household categories. Seed rate used by most sharecroppers is less than average, whereas seed rate used by lessee farmers is higher than average. No such concentrated trend was observed in seed rate used by other farmer categories.

In potato cultivation, the seed rate depends upon the season of planting, duration, seed size, spacing, and so on.[14] *Kufri Jyoti* and *Kufri Pukhraj* are both oval potatoes and the prescribed seed rate for oval potato is 13kg to 17kg per *katha*. However, on an average, farmers use 10kg potato seed per *katha* of land for TL seeds of first generation, and 15kg per *katha* for seed of second generation. Therefore, seed rate in potato cultivation is lower than the prescribed level. In mustard cultivation, prescribed seed rate is 30g to 50g per *katha*. Farmers of the two CD Blocks under study follow this prescribed seed rate.

Chemical Treatment of Seeds

Chemical treatment of seeds helps to protect germinating seeds and seedlings against pest attacks. It was found that regular chemical seed treatment is done by only a few farmers. Farmers responded that they perform chemical treatment of certified seeds with chemicals provided within the packet of seeds. Many seed companies also sell treated seeds nowadays. It has been found that only 13 farmer households out of 360 farmer households practice chemical treatment of TL paddy seeds during *aman* paddy and *boro* paddy season. Farmers informed that they treated the seeds chemically in years when other farmers of the village or their friends also did so. No one among the lessee and sharecroppers, who use farm saved seeds, follow chemical treatment of paddy seeds.

---

14. http://vikaspedia.in

The Quality of Seed

About 19.72 percent farmers opined that they have faced quality related problems with seeds at least one time in five years. Quality related problems faced by the farmers include low yield rate, admixture of seed varieties and poor seed germination. Among the different categories of farmer households, big farmer households have reported most frequently against seed quality. About 31.58 percent of the big farmer households reported that they have been cheated in buying seeds. It is the lessee farmer households who complained least against seed quality. Only 6.67 percent among the lessee farmer households faced low productivity due to low grade seeds. There is no such concrete explanation for such variation in reporting complains among farmer household categories. One probable cause is that since they have a large amount of cropped area, total production realized by a big or medium farmer falls considerably if seed quality deviates slightly; and thus they note it. However, lessees and sharecroppers generally cultivate a small amount of land and thus remain indifferent about seed quality unless they face a major problem caused by seeds.

Among the farmer households who reported about bad seed quality, more than 50 percent bought seeds from the market, about 24 percent bought seeds from government seed farms and other public sources, about 18.31 percent bought seeds from co-operatives, 5.63 percent bought seeds from local farmers and only 1.41 percent bought seeds from *aratdar* or *mahajan*. As the market is the major source of seeds to the farmers, so it is quite normal that most of the reports on bad quality seeds have come from farmers who bought seeds from the market. Farmers did not get any compensation or any kind of help from the seller or from the seed company. In contrast to this, there is an instance of compensation being provided by a co-operative society for bad quality paddy seeds. In 2015, Masjidpur co-operative society of Galsi-II CD Block replaced the paddy seeds which did not germinate properly, free of cost after farmers lodged a complaint.

Fertilizer

Fertilizer is an essential component of present-day agriculture. Though fertilizer may be organic or chemical, generally, it means chemical fertilizer. Change in fertilizer use is commonly equated with the HYV seed use, irrigation, fertilizer price and other agro-economic variables. However, Raju (1990) in her study, tried to show that extension network, supply and distribution network, and availability of credit to the farmers also determine fertilizer use. Raju (1990) conducted the study covering an agriculturally developed area and an under-developed area of Andhra Pradesh and observed that farmers of the less agriculturally developed area many times compelled to pay more for fertilizer and to depend on money lender. Therefore, it is highly needed to establish institutional support like the co-operative society for the efficient supply of inputs like fertilizer and curtailing their prices.

Chand and Pandey (2008) in their study, opined that fertilizer use experienced rapid growth during the period of the green revolution, but slowed down after 1996-97. It is also opined that fertilizer use in all of the Indian states is relatively low and there is a scope for production growth through more fertilizer use. Regarding NPK use, their study revealed that the use of N, P and K has moved towards the norm but still far away from the optimum level.

To be self-sufficient in food production, the Indian government provides subsidy for fertilizer to both, producers and farmers, so that fertilizer industry could supply fertilizer to the farmers and farmer could get it at affordable prices. There is a debate about whether the fertilizer subsidy benefits the farmers or the fertilizer industry (Sharma & Thaker, 2010). Moreover, studies on fertilizer use (Chand & Pandey, 2008; Sharma & Thaker, 2010) indicate that fertilizer use is high in a few states, and they enjoy the fertilizer subsidy more. However, a major issue related to fertilizer use at present is its increasing cost. A significant cause for the recent fertilizer price increase is the increase of global prices of raw material for phosphate and potash. The price of urea has not increased so much because of government control.

Sharma and Thaker (2010) are against subsidy reduction of fertilizer as they fear it would affect farm production and income of the small and marginal farmers. Regarding the mode of providing subsidy to the farmers, Sharma and Thaker (2010) stand against the direct transfer of subsidy to farmers as they fear this would result in misuse of subsidy. Instead, they favour the currently existing nutrient-based subsidy method.

Farmers use two types of fertilizers. One is bio-fertilizer and the other is chemical fertilizer. Before the Green Revolution in India, farmers mostly used to apply farm yard manure. However, use of chemical fertilizer started to increase with the initiation of the Green Revolution. In order to increase production, the government stated to provide subsidies to fertilizer industries so that farmers could get sufficient chemical fertilizers at lower cost. In West Bengal, farmers became accustomed to using more chemical fertilizers from the decade of the 1980s. Nowadays, we have become aware about the negative effects of chemical fertilizers on the environment as well as on human health. Besides, fertilizer prices are increasing day by day due to the global price rise of phosphate and potash. That is why chemical fertilizer use and its availability has become an important issue. We have examined the changes in fertilizer use, sources of fertilizers to the farmers, level of fertilizer use, and behavioural changes of farmers on fertilizer use in the study area.

Sources of Chemical Fertilizer

As per SAS 2003 reports, about 47 percent farmer households reported availability of fertilizer within one's own village during *kharif* season, whereas only about 27 percent farmer households from all over India reported such a facility during *kharif* season. In Galsi-I and Galsi-II CD Blocks, there are different sources of fertilizers available to farmers (Table-

3.5). The market is the major source of fertilizer, as more than 50 percent of the farmer households buy required fertilizers only from the market. Dependence on the market as the sole source of required fertilizer is highest among the lessee farmer households, and lowest among the medium farmer households. About 73.33 percent of the lessee farmer households are totally dependent on the market. Since farmers can pick up fertilizers of different branded companies from the market, they prefer to buy fertilizers from there if they have cash in hand. Moreover, many farmers have good relations with sellers, and can get fertilizers on credit.

On the other hand, cooperative societies sell fertilizers produced by IIFFCO (Indian Farmers Fertiliser Co-operative Limited) only. In addition, farmers, except members of that cooperative society, have to pay the bill of fertilizers purchased instantly to the authority. About 15.83 percent of the farmer households solely depend on co-operative societies. About 10.56 percent farmers purchase fertilizers from co-operative societies as well as from the market. Medium as well as big farmer households are most likely to enjoy this facility provided by co-operative societies-about 47.37 percent big farmer households and about 42.86 percent of the medium farmer households purchase fertilizer from co-operatives. As their fertilizer requirement is high, medium and big farmers prefer to buy fertilizer from co-operative because the price per packet is less than the market price, which helps to reduce the cost amount considerably. Other categories of farmer households are far behind in using co-operatives as a source of fertilizer. Sharecroppers rank last, followed by lessee cultivators in purchasing fertilizers from co-operatives. Only 10 percent of the sharecropper households and only 13.34 percent of the lessee farmer households buy fertilizers from co-operatives.

About 18 percent of the farmer households depend on small businessmen and *aratdar* for fertilizer. Farmers of the villages located far from the market and co-operative societies are also dependent on this source of fertilizer. Besides, poor

Table-3.5
Distribution of Farmer Households by Sources of Fertilizer at present

Present Source(s) of Fertilizer to the Farmers	Farmer Household (%)						
	Landless lessee	Landless Sharecropper	Marginal	Small	Medium	Big	All
Market/Fertilizer Dealer	73.33	55.00	53.19	53.95	38.10	42.11	51.94
Co-operative society	6.67	5.00	15.96	19.74	14.29	21.05	15.83
Local Small businessmen	0	15.00	5.32	3.95	4.76	5.26	5.28
Aratdar	13.33	20.00	14.89	9.21	9.52	5.26	12.78
Co-operative society and Market	0	0	9.04	10.53	23.81	26.32	11.11
Co-operative society and small businessmen	0	0	1.06	0	2.38	0	0.83
Co-operative and *Aratdar*	6.67	5.00	0	1.32	2.38	0	1.11
Market and small businessmen	0	0	0	1.32	2.38	0	0.56
Others	0	0	0.53	0	2.38	0	0.56
Total	**100**	**100**	**100**	**100**	**100**	**100**	**100**

Source: Field Survey, 2016-2018.

farmers, having deficit of cash in hand, also prefer to buy fertilizer from local small businessmen and/or *aratdar* as they provide it on credit. Though the *aratdar* and local small businessmen appear more or less same as a seller of fertilizer, the main difference between them is that *aratdar* provides fertilizers and sometimes other inputs on credit, and in turn sets pre-conditions of selling produce to them. In many villages it is observed that grocers act as local suppliers of fertilizer to the farmers during the cropping season. They have also been considered as local small businessmen here.

The amount required by the farmers and urgency of the fertilizer requirement also determines which sources are used. For example, if a farmer needs less amount of fertilizer and/ or has no time to go the market/co-operative, he purchases it from local small businessmen or from *aratdar*. Personal relationship is another factor. Some of the farmers, even medium and big farmers, buy fertilizers from local businessmen because of good interpersonal relationships.

Sources of Fertilizer Before 1995

Sources of chemical fertilizer to the farmer households before 1995 have been shown in Table-3.6. Farmer households who failed to respond about the past source of chemical fertilizer have been excluded from these calculations. The market was the main source of chemical fertilizers to the farmers even before 1995, though dependency on the market was much lower compared to the present situation. Moreover, dependency on *aratdar* for chemical fertilizer was higher than at present. Earlier, about 21 percent farmer households were solely dependent on *aratdar* for chemical fertilizer. That means dependence on *aratdar* has declined over time and it certainly indicates improvement in the village economy. However, the role of co-operative society as a sole supplier of chemical fertilizers to farmers has also slightly declined over time. This decline may be an indicator of deterioration of co-operative societies. Most of the farmers used to purchase the required chemical fertilizers from only

Table-3.6
Distribution of Farmer Households by Sources of Fertilizer before 1995

Sources of Fertilizer to the Farmers	Farmer Household (%)						
	Landless lessee	Landless Sharecropper	Marginal	Small	Medium	Big	All
Market	53.33	22.22	45.51	53.62	42.5	40.00	45.68
Co-operative society	0	0	17.37	14.49	25.00	33.33	16.67
Local Small businessmen	33.33	22.22	7.19	7.25	10.00	13.33	9.88
Co-operative society and market	0	0	3.59	4.35	10.00	6.67	4.32
Co-operative society and Aratdar	0	0	1.20	0	0	0	0.62
Aratdar	13.33	44.44	22.75	20.29	12.5	6.67	20.99
Landowner/other farmers	0	11.11	2.40	0	0	0	1.85
	100	100	100	100	100	100	100

Source: Field Survey, 2016-2018.

one of the sources.

The main source of chemical fertilizer to the sharecroppers was the *aratdar*. About 44.44 percent sharecropper households used to purchase chemical fertilizer from *aratdar*. Among lessee farmer households, about 13.33 percent depended on *aratdar* for chemical fertilizer. Local small businessmen were the second important source for the lessee farmers and sharecroppers. Only about 11.11 percent of the sharecropper households were dependent on landowner/other farmers for fertilizers. Role of co-operative societies as a supplier of fertilizer to the lessee or sharecropper households was nil.

About 45.51 percent of the marginal and 53.62 percent of small farmer households used to purchase chemical fertilizer from the market. *Aratdar* was the second important source of chemical fertilizer to these farmer household categories. Co-operative society ranked third in chemical fertilizer supply to the marginal and small farmers. However, it was the second most important source of chemical fertilizers to the medium and big farmers. These farmer household categories were most privileged in enjoying services provided by co-operative societies. More than one-third of the medium and big farmer households grabbed the scope of buying fertilizer from this source.

Therefore, analyzing the sources of fertilizer to farmers in the past as well as at present, it can be concluded that dependence of sharecroppers, marginal and small farmers on *aratdar* for fertilizer has declined considerably over time. Lessee farmers, sharecroppers, small and marginal farmers, to some extent, succeed in getting these advantages from co-operative societies at present, though medium and big farmers benefit the most.

Change in Fertilizer Use between 1995 and 2018

The two CD Blocks under study have sufficient supply of fertilizer. However, supply is not the only variable that determines fertilizer use. Resource endowment, use of HYV seeds, fertilizer prices, output prices and extension services are the other variables which exert influence on fertilizer use in a particular region (Raju, 1990; Sharma & Thaker, 2011). Moreover, optimum use of fertilizer is very important to get highest output as well as to keep soil quality intact and to make agriculture sustainable.

Farmers of the study area generally use DAP, urea and potash in *aman* and *boro* paddy cultivation. During *boro* paddy cultivation, many farmers also use NPK 10:26:26 in place of potash. In potato cultivation NPK 10:26:26 is the most used fertilizer. In this section, use of chemical fertilizer in terms of N, P and K for paddy cultivation has been discussed. Normal ranges of these nutrients use have been determined on the basis of the fertilizer chart provided by ADA office of Galsi-I CD Block (Appendix 3.1). There is similarity in fertilizer use by the farmers of both CD Blocks. Therefore, combined data of both CD Blocks have been presented.

Table-3.7 shows that nitrogen use is high among the farmer households. About 54 percent of the farmer households apply nitrogen more than the recommended level. Only about 19.58 percent farmer households use nitrogen within the prescribed range. Among different farmer household categories, lessees rank first in high nitrogen use followed by small, sharecroppers and marginal farmer households. However, most of the big farmer households (52.94%) use nitrogen within normal range. DAP is the main source of phosphate and its use is also very high. More than 85 percent of the farmer households use phosphate more than the recommended level. Only 6.53 percent farmer households use phosphate within the recommended level. Unlike nitrogen and phosphate use, a very high percentage of farmers use potash below the normal level. About 42.73 percent farmer households use potash below normal range. High amount of urea fertilizer use is mainly responsible for the high amount of nitrogen use in *aman* paddy cultivation.

Resource-poor farmers use more urea because the rate

Table-3.7

Distribution of Farmer Households by the Use of NPK during Kharif Paddy Cultivation

Nutrients		Farmer Household (%)						
		Landless lessee	Landless Sharecropper	Marginal	Small	Medium	Big	All
Nitrogen Use	Below Normal	16.67	31.25	30.22	20.00	25.71	17.65	26.41
	Normal (11-12Kg)	16.67	12.5	14.84	20.00	31.43	52.94	19.58
	Above	66.67	56.25	54.95	60.00	42.86	29.41	54.01
Phosphate Use	Below Normal	8.33	6.25	10.99	4.00	5.71	5.88	8.31
	Normal (5-6Kg)	16.67	6.25	5.49	8.00	5.71	5.88	6.53
	Above	75.00	87.5	83.52	88.00	88.57	88.24	85.16
Potash Use	Below Normal	50.00	56.25	42.86	42.67	40.00	29.41	42.73
	Normal (5-6Kg)	41.67	18.75	28.57	28.00	28.57	41.18	29.08
	Above	8.33	25.00	28.57	29.33	31.43	29.41	28.19

Source: Field Survey, 2016-2018; normal range is calculated per bigha.

of urea is cheaper than DAP and potash. Rate of urea was only Rs. 8 per kg, whereas rates of DAP and potash per kg in the open market were Rs. 17 and Rs. 30 respectively, on an average during our survey. Except for the price difference, lack of proper knowledge regarding soil nutrients and its use also determines fertilizer use. Many farmers told us that potash retarded the growth of plants and it is not very helpful for increasing production. However, the fact is that potash is also an essential nutrient for plant growth because it helps to enhance crop resistance to pests and diseases, improve root growth and increase weight of the paddy. Most farmers use more urea as it helps plants to grow rapidly. However, in spite of having better financial condition, big farmers use less amount of urea probably because they have more awareness of the negative impacts of nitrogen than the other categories of farmers.

In the study area chemical fertilizer use is increasing day by day. Thus, farmers were asked to comment on the main cause of growing chemical fertilizer use. About 34.17 percent farmers did not respond to this question, whereas only 2.2 percent farmers opined that no changes had taken place in last 20 years. However, about 26.39 percent of the farmer households answered that soil fertility is decreasing and that is why they are applying more and more fertilizer to keep the production intact over time, while 23.69 percent of the farmer households responded that they apply more fertilizer aiming at more crop production. About 9.17 percent households answered that soil fertility is decreasing while household expenditure is increasing, as a result of which they are using more fertilizer to attain more output from agriculture. Only 4.4 percent farmer households talked about declining bio-fertilizer use. According to them, they are being compelled to use more chemical fertilizer because of soil fertility loss caused by declining use of bio-fertilizer.

However, manure use has declined in Galsi-I and Galsi-II CD Blocks as in the rest of West Bengal. From the data published by the Directorate of Economics and Statistics, it is observed that in West Bengal manure use has declined from 40.16 quintal per hectare in 1996-1997 to 20.37 quintal per hectare in 2015-16. Farmers of all sample villages opined that almost every farmer household used to apply manure every year in their agricultural land before. However, at present the supply of manure has become limited due to declining cattle rearing (causes for limited supply of manure have been discussed in Chapter 6). Declining use of *ghani* oil made from self-produced mustard, and rising price of mustard oil cake is also responsible for declining manure use. It is observed that only about 51.39 percent of the total farmer households surveyed use manure. In Galsi-II CD Block, larger number of farmer households use manure than in Galsi-I CD Block. Out of 18 sample villages, at 5 villages less than or equal to 25 percent farmer households use manure. Manure use is high among the farmer households of Dalimgaria, Sankrai, Taherpur and Adrahati (Appendix 3.2), where more than 75 percent of the farmer households use manure.

Use of manure is least among sharecroppers and highest among the big farmer households (Fig. 3.1). About 59 percent big farmer households use manure. Percentage of

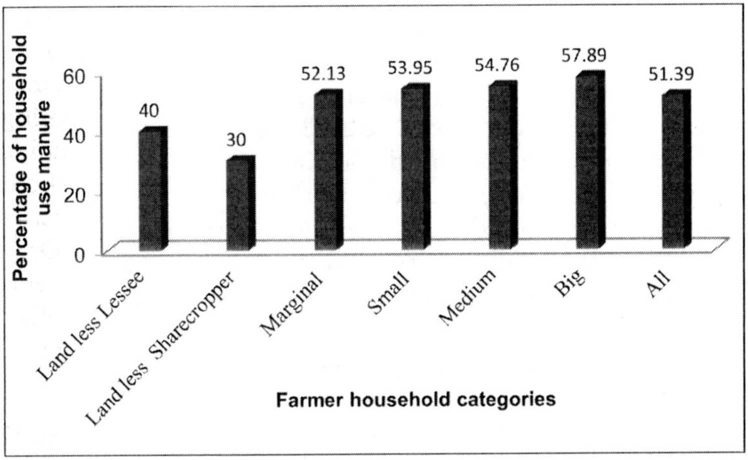

Fig. 3.1: Category-wise Distribution of Farmer Household by Manure Use

Source: Field Survey, 2016-2018

sharecropper farmer households using manure is only 30 percent. Use of manure among the lessee farmer households is also very low. Only 40 percent of them use manure. As cow dung manure is the main source of manure, generally farmer households having cows use manure. However, in spite of having cows in more than 73 percent of the lessee households, manure is used by only 40 percent of them. Use of cow dung as fuel is one cause for such behaviour. Besides, there is no guarantee of getting the land on lease, and thus they do not keep cow dung for manure. They like to use chemical fertilizer only.

The picture of manure use becomes dark if one looks into the amount and interval of manure use. Farmers admitted that in past days they used to apply ample manure at the beginning of every crop season. Many of them also used oil cake as fertilizer in paddy cultivation. However, at present most of the farmers apply manure insufficiently in some portions of their land with an interval of one year or even two years. Only 19 out of 360 farmer households use oil cake manure at present.

Use of Chemical Pesticides and Plant Protection Chemicals

Like fertilizer, chemical pesticide is also an important input of agriculture. It helps to secure production by controlling pest and diseases. Thus, availability of pesticide to the farmers is an important issue. As per SAS 2003 report, about 44 percent of farmer households reported availability of pesticides in their own village during *kharif* season, and about 97.7 percent farmer households received adequate pesticides during *kharif* season in West Bengal. Availability of pesticides was higher in West Bengal than the national level average. In India, on an average, only about 19 percent of farmer households reported availability of pesticides in their own village. However, farmers all over India got adequate supply of pesticide during *kharif* season-about 96.1 percent farmer households all over India received adequate pesticides during *kharif* season.

Farmers purchase the required pesticides from retail

shops. None of the farmers reported lack of pesticides when needed. Rather, pesticides of different companies are available to the farmers, so sometimes they become confused in choosing which brand to use. In this connection, it is important to note that most of the farmers do not know the composition of pesticides. Most of them repeatedly use pesticides of those brands which provide satisfactory results in pest and disease control quickly. Besides, farmers also use new brands at the recommendation of retailers. Role of ADA officials and staff in guiding farmers about pesticide use is almost zero.

The most-used pesticides in paddy cultivation are Asataf (for controlling sucking and chewing insects), Glamore (for controlling hopper burn in later crop stages), DuPont Ferterra (used for long term protection against early shoot borer and top borer), Contaf Plus (effective for controlling rust and leaf spot, powdery mildews and Rice Sheath Blight). Miraculan is the most used plant growth regulator in the study area.

Farmers stated that chemical pesticide use has increased over time. However, about 33.61 percent farmer households failed to provide cause(s) for such an increasing trend in chemical pesticide use. From the answers provided by the rest of the farmers, it seems that growing intensity of pest attacks, increasing resistance power of pests to the chemical pesticides and reduced effectiveness of pesticides are the major causes for growing chemical pesticide use. Farmers have no experience with the use of bio-pesticides. During focus group discussion at Kurkuba village of Galsi-II Block, it was found that many years ago, farmers tried to control pests following integrated pest management (IPM) guided by the ADA officials, but failed to control the pests due to the absence of those officials during *puja* vacations, because of which farmers faced huge losses. After that, they never tried such methods of pest control.

Irrigation

Present-day Indian agriculture is heavily dependent on

irrigation. However, irrigation water is becoming a scarce resource. Therefore, it is needed to know about access to irrigation water and its management. Pingle (2011) in his study on the past and present day's irrigation practices in Telangana, has stated that tank irrigation in Telangana has a long history and it is best-suited method of irrigation in this region, however, despite higher costs than that of tank irrigation, area irrigated by tubewells has grown. Pingle (2011) opined that lack of fund for tank irrigation, negligence of political-administrative class toward tank irrigation, the encroachment of tank beds and low amount of rainfall are the leading causes of the decline in tank irrigation while tubewell irrigation is increasing due to daily seven hours of power supply free of cost.

Ray and Ghosh (2007) have also experienced similar changes in districts of West Bengal. They showed that being a summer crop, *boro* paddy is dependent on irrigation of which groundwater contributes 77.11 percent, and area under groundwater irrigation is increasing at the expense of area under the tank and canal irrigation. Due to the excessive use of groundwater, ecological problems are emerging.

In these studies, they have only shown the changes in irrigation systems but do not mention the techniques to save water. This type of study has been done by Thakur et al., (2011). Thakur et al. mentioned different methods to save water in rice cultivation. According to them, technologies such as field channel construction, land leveling, tillage and puddling, bund preparation and use of raised beds are useful to save water during land preparation and before main crop establishment, whereas Saturated Soil Culture and alternating wetting and drying methods are useful during the crop growth period. Palanisami et al., (2011), based on primary as well as secondary data and using different statistical techniques like regression analysis, ANOVA etc., have made a study on the spread of micro irrigation (MI) in India. Palanisami et al. (2011) depicted that there are spatial variations among farmer categories adopting MI. According to them, the adoption rate of MI technology is low enough compared to potential because of technological complexity and other socio-economic constraints. It is also opined that uniform subsidy across the state, modifications in agro techniques like paired row planting and training of farmers will be helpful in spreading MI system all over India.

Sarkar and Das (2014) opined that shift from canal irrigation to tubewell irrigation in Punjab with the help of free supply of electricity resulted in groundwater depletion as well as agro-ecologically misfit crop selection. Due to foodgrain procurement policy and free supply of electricity for agriculture, paddy production is more profitable to the farmers, and thus the most recommended practice of crop diversification has remained unsuccessful.

Sources of Irrigation

In terms of irrigation infrastructure, Purba Bardhaman district is better developed in comparison to other districts of West Bengal. Samanta (2001, p.85) has noted, "During the Mughal period, the region had a well-developed indigenous system of overflow irrigation. The introduction of modern canal irrigation took place in the region in 1881 with the construction of Eden Canal". From the 1970s, shallow and submersible irrigation systems started to develop in this district, which resulted in expansion of *boro* paddy cultivation. At present, canal and submersible/shallow tubewells are the two important sources of irrigation.

Among the different sources of public irrigation, the canal network occupies the prime position with 90.36 percent of the total irrigated area in the Purba Bardhaman district. River lift irrigation has lost significance and provides for only 3.59 percent of the total irrigated area. Deep tubewells irrigate only 6.05 percent of the irrigated area (Bureau of Applied Economics and Statistics, 2014). Tank irrigation and shallow tubewells have lost their importance as sources of irrigation. At the block level, the share of canal irrigation varies between 0 and 99.89 percent. From the percentage share of different sources irrigation, it is quite clear that the public irrigation infrastructure of the region is still dominated by canal irrigation.

Farmers' Concerns over Irrigation System

Members of farmer households were asked questions about the changes in irrigation systems in the two decades after 1990s. About 40 percent of the farmer households opined that irrigation system remained the same as it was before 1995 (Table-3.8). About 33.06 percent farmer households believe the irrigation system has improved, but about 27 percent farmer households reported worsening of irrigation systems. Though there are variations in opinion among farmer household categories, it is quite impossible to provide any general explanation for such variation. Therefore, it needs further investigation.

Table-3.8
Farmer Households' Opinion on Irrigation System Development

Farmer households categories	Farmer households opined irrigation system has		
	Improved	Worsen	Remained same
Landless lessee	40	20	40
Landless sharecropper	45	15	40
Marginal	36.17	26.60	37.23
Small	28.95	31.58	39.47
Medium	26.19	26.19	47.62
Big	15.79	31.58	52.63
All	33.06	26.94	40.00

Source: Field Survey, 2016-2018.

Though most of the agricultural land area is irrigated by canal, only 35.59 percent farmer households are satisfied with the canal irrigation system at present (Table-3.9). Farmer households of Galsi-II CD Block are less satisfied than the farmer households of Galsi-I CD Block. Therefore, low level of satisfaction regarding canal irrigation system indicates that there are certain problems in this system.

Table-3.9
Category-wise Distribution of Farmer Households Satisfied by Canal Irrigation

Farmer house-hold categories	Satisfied farmer households (%)		
	Galsi-I	Galsi-II	Combined
Landless Lessee	54.55	50.00	53.33
Landless Sharecropper	33.33	57.14	42.11
Marginal	36.56	45.12	40.57
Small	18.92	30.30	24.29
Medium	44.44	16.67	28.57
Big	11.11	40.00	26.32
All	33.33	38.13	35.59

Source: Field Survey, 2016-2018.

The major problems of the existing canal irrigation system as perceived by the farmer households have been listed in Table-3.10. From this table, it is observed that unavailability of canal water in every *boro* season for paddy cultivation is the most important problem, as it was pointed out by most farmer households. In some parts of these CD Blocks, there is a quota system. These portions have been getting canal water alternate years for *boro* paddy cultivation for the last 18 years or more. About 28 percent farmer households reported canal water for irrigation was available before, but not at present. Moreover, farmers also reported that they get irregular and insufficient supply. About 12.65 percent farmer households reported that crop loss occurred during monsoon season duc to excessive canal water supply. All these problems faced by the farmers regarding canal water indicate lack of maintenance of the canals and the declining water storage capacity of the reservoirs. Various newspaper reports also support this claim ("Chaser jol niye kshov chasider", 2015; "Milche na secher jol", 2015). The system of collecting tax from farmers for canal irrigation has also deteriorated. Therefore, it can be said that the existing irrigation system, based on which tremendous development in paddy cultivation took place before, is a threat to the farmers at present. This system is losing its efficiency day by day. Large scale investment is required for the renovation of the canal network system in the study area.

Table-3.10
Problems of Canal Irrigation in Galsi-I and Galsi-II CD Blocks

S.No.	Problems	Farmer households
1.	Irrigation water was available before but is not at present	27.94
2.	Irrigation water is not available for every boro season	45.00
3.	Amount of water supplied is insufficient	27.94
4.	Long time interval of providing irrigation water within one crop season	34.12
5.	Do not provide water in time	27.94
6.	Faced loss due to excess water supply during monsoon	12.65

Source: Field Survey, 2016-2018.

As mentioned before, groundwater is another major source of irrigation. By and large, groundwater is a supplementary source of irrigation to canal irrigation. However, in some parts of the study area groundwater is the main source of irrigation. For example, Taherpur and surrounding villages have no area under canal irrigation. The government has set submersibles in some places. Taherpur village of Galsi-II CD Block has two submersibles, run in a co-operative system. However, most of the submersibles have been set up by individual farmers.

Distribution of farmer households having submersibles in the surveyed villages (Table-3.11) reveals that 65 percent of the surveyed farmer households at Mallasarul of Galsi-I CD Block have submersibles, and 50 percent of the surveyed farmer households at Sanota village of Galsi-II CD Block possess submersibles. In 7 out of 18 villages, none of the surveyed farmer households possess submersibles. Therefore, it is observed that submersibles are unevenly distributed.

Table-3.11
Distribution of Farmer Households having Submersibles under Sample Villages

Sample villages of Galsi-I CD Block	No. of Farmer households	Sample villages of Galsi-II CD Block	No. of Farmer households
Kolkol	0	Kurkuba	0
Ramgopalpur	3	Sasanga	1
Jagulipara	3	Sathinandi	2
Bhirsin	6	Saonta	12
Mollasarul	13	Sankrai	2
Dwarnari	0	Dayalpur	0
Chaktentul	0	Taherpur	3
Mankar	0	Dalimgaria	2
Bandutia	0	Adrahati	2
Total	**25**	**Total**	**24**

Source: Field Survey, 2016-2018.

Distribution of shallow tubewells or submersibles among the farmer households (Fig. 3.2) depicts that about 61 percent owners of shallow tubewells/submersibles are medium and big farmers. Therefore, most of the resource-poor farmers are compelled to depend on the submersible owners who are

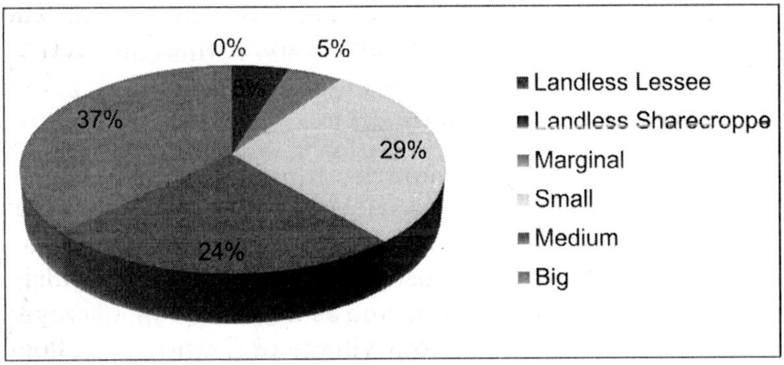

Fig. 3.2: Distribution of Shallow Tubewell/Submersible among Farmer Household

Source: Field Survey, 2016-2018.

basically from the affluent section of society. Submersible owners make a profit either by leasing in land under their own submersible's command area or/and by selling irrigation water to the other farmers. Besides, some submersible owners enjoy subsidy for electricity consumption by submersibles. However, due to the increasing price of electricity or diesel, resource-poor farmers are suffering the most.

The declining water table is also another concern for groundwater irrigation. Farmers opined that during the 1980s and 1990s, many shallow tubewells were in operation. However, due to decline in the water table caused by excessive mining of groundwater, most of the shallow tubewells are obsolete at present and have been replaced by submersibles. Irrigation using submersibles is more costly than using shallow tubewells. Due to rising cost, *boro* paddy cultivation with the help of privately purchased irrigation water is not profitable enough for many farmers. They are therefore searching for alternatives.

Labour

Paddy is a labour intensive crop. Farmers generally hire labour for land preparation, transplanting of paddy saplings, crop harvesting, and so on. Rogaly (1999) in his study related to seasonal migration and agrarian change in West Bengal opined that rapid rice output growth in West Bengal during the 1980s and early 1990s was associated with demand for hired labour by smallholder landowners. In this context, we would like to briefly discuss this input and show the changes related to labour use in paddy cultivation.

Like other developing countries, Indian agriculture is also labour intensive. Innumerable studies have been conducted on the migration of labour, labour crisis, attached labour etc. Rogaly (1999) investigates the relationship between seasonal migration of agricultural workers and changes in agrarian structures in West Bengal. Rogaly opined that rapid growth in the output of rice in West Bengal, in general, and *aman* paddy, in particular during the 1980s and 1990s, was associated with the increase in the quantity of hired labour demanded by smallholder landowners. *Aman* paddy cultivation in Purba Bardhaman and Hugli in West Bengal was dependent on the migrant labour coming from various districts of West Bengal and Bihar. Poorer, the labour-selling sections of these areas such as Santal, Bhumij, Muslim-bhumij migrate to Bardhaman and Hugli as in these areas wage rate is high and as they expect to return with a lump sum of cash. He has also explained the social change due to labour migration and the problems faced by the migrants. His concluding remark is that seasonal migration in West Bengal is less exploitative and less, dangerous but potentially progressive.

Jodhka (2012), based on primary data collected in 1988 and 2008-09, has studied on attached labour in Haryana. For this study, the author has selected two villages of Karnal district in Haryana. Jodhka has dicussed the changing class relations in rural areas with a special focus on attached and un-free labours. His study suggests that the number of people working on the land is declining. Besides, the demand for labour is also declining. Casual and contractual labour have replaced attached labour, though big farmers still employ attached labours.

Labour Use and Its Change Over Time

Galsi-I and Galsi-II CD Blocks have a high proportion of agricultural labour. Therefore, farmers used to hire local labour for agricultural activities along with using their own household labour. However, since paddy cultivation creates very sharp seasonal peak in labour demand, the local labour supply was insufficient to meet the demand and thus farmers needed to hire labour from outside. Seasonal migration of agricultural workers was a very common element of the agrarian structure of Purba Bardhaman district. In-migration of agricultural labour from Malda, Murshidabad, Bankura, Purulia and even from Bihar helped farmers to meet their labour demand. This type of in-migration was more pronounced during the *boro* paddy cultivation. Due to availability of local as well as migrant labour, there was no crisis of labour. In most cases, they were paid on a contractual basis. On the other hand, local labourers were used to receiving pay on a daily basis. Farmers opined that local as well as outside labourers used to work hard and they were loyal enough. The wage was relatively lower, but for the migrant labour it was more remunerative than their native places. Though wages of migrant labour were higher, they were more productive and did not object to working longer days; thus employer farmers preferred them more. Rogaly (1999) highlighted some instances of exploitation of the migrant labourers, but such migration was helpful for both parties-labourers and employers.

However, many changes have taken place in labour use since then. The intensity of agricultural labour migration has reduced considerably. One of the major causes, as perceived by the farmers, is decline in area under *boro* paddy cultivation as well as low profit margin, barring farmers from employing high-cost migrant labour. Harvester use is also responsible for reducing migrant agricultural labour. As a result, dependence on local labour has increased considerably, resulting in better bargaining power of the local labour. Farmers accused that agricultural labourers either do not want to work on a daily basis or work very less under this agreement. That is why a contractual labour arrangement has become popular. Under this agreement, agricultural labourers complete the stipulated work as per agreement and are paid by the employers either in cash or a pre-determined amount of crop and cash. Terms and conditions under this agreement vary from place to place. Mohanty (1999, p.A58) opined that contract labour is "now widely preferred among the landowners because it helps meet deadlines for operations". Working of daily labour may also be considered one kind of contract because a daily labourer also demands wages based on amount of work, not based on time.

The single most important difference between the present day daily labour and so-called contract labour is that contract labourers work in a group and try to finish the work as quickly as possible, but daily labourers may work individually and may work less hurriedly. A daily labourer earns a full wage of about Rs. 180-200 by detaching seedlings from the bed and transplanting them, amounting to 32-40 bundles of paddy seedlings, which is more or less equivalent to transplantation of paddy covering 2.5 *katha* of land. A daily labourer normally takes 5-6 hours to complete this work. Farmers accused that nowadays labourers are less loyal and work less compared to the past. For paddy cutting and binding, farmers had to pay wages of three labourers, varying between Rs. 1200-1600 per bigha. A group of six to eight daily labourers took about 16 hours to complete the work, over a period of two days. Local agricultural labourers do not like to be engaged in paddy transplantation during the second half of the day. Such changes in labour use and its nature indicate a change in the power relations between landowner farmers and agricultural labourers. Due to the labour crisis for agricultural activities, agricultural labourers enjoy more power in determining the wage rate and terms and conditions of working.

Labour Supply

Insufficient labour supply for agricultural work during the peak season of transplanting of paddy or harvesting is another issue. As per Census of 2011, more than 50 percent of the total workers in Galsi-I and II CD Bocks are agricultural labourers. However, about 54 percent and about 53 percent of the farmer households reported unavailability of required labour during *aman* and *boro* paddy cultivation respectively (Table-3.12). Social change is the main cause for unavailability of labour at present. Due to spread of education and modernity, the new generation of agricultural labour households do not like to be engaged in agricultural activity, and rather prefer to be engaged in non-farm activities in urban and industrial areas. Financial benefit is also a reason, as they can get more working days and sometimes earn more from these non-farm activities. Out-migration of moderately or less-educated young male members of the labour households to nearby urban centres or to other states is a common phenomenon in the study area.

Table-3.12
Farmer Households' Opinion on Availability of Labour

Response		Percentage of Farmer Household						
		Landless lessee	Landless Sharecropper	Marginal	Small	Medium	Big	All
Aman	Available	40	40	39.36	42.86	53.66	42.11	41.94
	Unavailable	33.33	45	57.45	57.14	43.9	52.63	53.89
	No response	26.67	15	3.19	0	2.44	5.26	4.17
	Total	**100**	**100**	**100**	**100**	**100**	**100**	**100**
Boro	Available	44.44	42.86	40.3	43.1	48.15	33.33	41.83
	Unavailable	33.33	35.71	54.48	56.9	48.15	66.67	52.99
	No response	22.22	21.43	5.22	0	3.7	0	5.18
	Total	**100**	**100**	**100**	**100**	**100**	**100**	**100**

Source: Field Survey, 2016-2018.

A low percentage of the landless lessees and landless sharecroppers reported about unavailability of agricultural labour, because they depend less on hired labour and rely more on household labour force. It is the small and marginal farmer households who reported about the lack of labour in larger numbers. Most of them opined that since medium or big farmers have more landholdings, the agricultural labourers expect to get more working days without any break from them, and thus these farmer household categories face less labour crisis. Besides, rich farmer households pay a little more wage or sometimes pay the agricultural labourers in advance. That is why resource poor farmers face more problems at the time of labour crisis. Another important aspect of labour availability is that male labour for agricultural work is rarely available.

Harvester use has helped a lot in reducing labour demand for paddy harvesting recently. However, labour demand remained high for other activities in paddy cultivation and for other types of crop cultivation. Many farmer households opined that they lease out their agricultural land due to lack of labour.

Impact of MGNREGS on Labour Use

MGNREGS has also had an immense impact on labour use and its availability. This scheme has certainly help to raise the income of the rural people. This scheme has also helped to increase wage rate. However, most of the farmer households accused that MGNREGS is responsible for the agrarian labour crisis, as works under this scheme are performed even during the season of peak demand for agrarian labour. Farmers also blamed that working hours and efficiency has decreased considerably and work ethics have been destroyed due to the implementation of this MGNREG scheme. Impacts of the MGNREGS as perceived by the farmer households have been presented in Table-3.13. Only about 4 percent farmer households opined that MGNREGS has had no effect on labour use. About 5 percent farmer households declined to respond. However, about 40 percent farmer households opined that this scheme is responsible for the labour crisis, wage increase, as well as working hour decrease. However, farmers of some villages informed that it has been decided to stop works under MGNREGS during peak demand period of labour for agriculture so that agricultural activities would not hamper.

However, a conflict has developed in many villages between landowner farmers and agricultural workers due to machine use and MGNREGS. Farmers blamed that MGNREGS has had a huge impact on labour cost and labour availability. On the other hand, agricultural labourers asserted that due to machine use in paddy harvesting, they have lost considerable amount of income and thus they have decided to further increase the wage. It has been perceived that harvester use has reduced labour demand and weakened the bargaining power of the agricultural labour to some extent. Farmers are also eager to use more machine labour to increase profit from cultivation by reducing labour costs.

Farm Implements as Input

Various farm implements meet different agricultural purposes. Data collected from the field survey (Table-3.14) reveal that possession of farm implements by farmer households, such as animal-drawn plough, animal-drawn cart and *duni* (traditional implement used for tank irrigation) has decreased over time. On the other hand, possession of farm implements such as spray machine, thresher, pump and tractor has increased.

However, very few farmer households among the landless lessee, landless sharecropper and marginal farmer households possess implements like spray machines, pumps, threshers and tractors. Most of the big farmer households possess pumps and tractors. As a result, farmer households that do not possess these have to hire them from other farmers. It adds extra cost to crop cultivation. For example, a farmer has to pay Rs. 30 to Rs. 60 per day for hiring a spray machine. Rent for hiring a man-powered paddy thresher ranges between Rs. 120 to 200 per day. Rate of pump for irrigating a land varies between Rs.

Table-3.13

Impacts of MGNREGS on Labour Use as Perceived by the Farmer Households

Impacts Fertilizer to the Farmers	Percentage of Farmer Household						
	Landless lessee	Landless Sharecropper	Marginal	Small	Medium	Big	All
1. Labour supply decreased	0	0	2.66	3.95	2.38	0	2.5
2. Wage increased	13.33	0	5.32	6.58	4.76	10.53	5.83
3. Working hours decreased	13.33	15	4.79	6.58	2.38	5.26	5.83
4. All of the above	33.33	15	41.49	44.74	42.86	26.32	39.72
5. Labour supply decreased and wage increased	0	10	3.72	6.58	16.67	0	5.83
6. Labour supply and working hours decreased	0	0	1.6	1.32	2.38	15.79	2.22
7. Wage increased but working hours decreased	26.67	45	28.72	26.32	26.19	31.58	28.89
8. No impact	0	10	4.79	2.63	2.38	0	3.89
9. No response	13.33	5	6.91	1.32	0	10.53	5.28
Total	**100**	**100**	**100**	**100**	**100**	**100**	**100**

Source: Field Survey, 2016-2018.

Table-3.14

Distribution of Farmer Households by Possession of Different Farm Implements

	Percentage of Farmer Household						
	Landless lessee	**Landless Sharecropper**	**Marginal**	**Small**	**Medium**	**Big**	**All**
a) Animal drawn Plough							
Before 1995	13.33	20	59.57	88.16	83.33	84.21	65.56
At present (2016)	6.67	10	9.04	18.42	19.05	21.05	12.78
b) Animal drawn cart							
Before 1995	0	15	45.21	71.05	66.67	78.95	51.39
At present (2016)	6.67	10.00	9.04	17.11	11.90	21.05	11.67
c) Duni							
Before 1995	13.33	20.00	43.62	69.74	73.81	57.89	50.83
At present (2016)	0	0	0	1.06	2.63	2.38	1.39
d) Spray Machine							
Before 1995	6.67	15.00	42.02	65.79	66.67	100.0	50.00
At present (2016)	40.00	25.00	61.17	94.74	80.95	100.0	69.72
e) Thresher							
Before 1995	6.67	5.00	45.74	67.11	73.81	89.47	51.94
At present (2016)	20.00	25.00	64.36	93.42	90.48	94.74	71.11

f) Pump							
Before 1995	6.67	5.00	18.09	31.58	42.86	47.37	24.17
At present (2016)	6.67	10.00	22.34	52.63	57.14	73.68	34.17
g) Tractor							
Before 1995	0	0	1.60	9.21	4.76	21.05	4.44
At present (2016)	0	0	4.26	17.11	23.81	57.89	11.67

Source: Field Survey, 2016-2018.

100 to 150 per hour. Ploughing of land using a rotavator cost Rs. 900 to 1000 per hour. Though farmers enjoy many advantages from using these machines, it is the owners of the implements who enjoy more advantages as they make a profit by hiring out their farm implements. Contrary to this, resource poor farmer households who lack the money to buy these modern implements are losing their self-sufficiency. There are other issues related to the technology trap as well, which we discuss in Chapter 6.

Conclusion

This chapter reveals that many changes in input use have taken place during the last twenty years. Not all these changes have proved to be good for agriculture or for the farmers. Small and marginal farmers have become more dependent on the market as well as on affluent farmers for agricultural inputs. Due to poor financial conditions, they still lag behind in reaping the benefits of modern agriculture. Labour crisis and inadequate canal water supply for irrigation are the most significant issues related to inputs in the study area.

Positive aspects related to input use in the study area are: 1) there is adequate supply of seed and fertilizer; 2) use of chemical fertilizer and pesticides is not too high; 3) farmers still, though few in number, use manure; 4) ADA offices are providing training to the farmers on how to reduce input use; and 5) some of the co-operative societies, which provide seeds, fertilizers and financial support, are still functional and also provide some support to the farmers. Having CRSMF in Bardhaman and KVK at Budbud, there is scope for getting some support from experts for agricultural development. If the DVC canal irrigation system is renovated, problems related to the irrigation system will be solved to some extent. As this chapter highlights various issues related to inputs in agriculture, except input costs, issues related to costs and profits from cultivation have been addressed in the next chapter.

4 The Profit: Actual and Perceived by Farmers

Introduction

The agrarian crisis is one of the most critical issues in India nowadays (Raghavan, 2008; Posani, 2009; Shroff, 2019). Farmers' suicide has also become a common phenomenon after the early 1990s (Narayanamoorthy, 2013, p.104). It is not only confined in backward regions, instead farmers of agriculturally developed regions such as Maharashtra, Karnataka, West Bengal, Andhra Pradesh and Punjab are also commiting suicide mainly because of indebtedness (Biru & Barpujari, n.d.). In this connection, it is opined that "indebtedness and other related problems are mainly caused by poor returns from crop cultivation; therefore, one should study the issue of profitability" (Narayanamoorthy, 2013, p.106; Pushpa, Srivastava, & Agarwal, 2017, p.389).

Before the Green Revolution in India, the main problems facing the agricultural sector were insufficient crop production and lower yield rate (Rao, 2004, p.20; Srinivasan, 2007, p.131). At that time, the primary concern of the agricultural price policy was to make sure that the difference between demand and supply of food crops did not result in too much price rise and that minimum support price (MSP) of the food crops were below market price (Ray, 2007, p.36; Tripathi, 2013, p.91). However, meager profitability and frequent loss in crop production is the main concern at present. Though the current price policy aims to protect the farmers' income and to provide them incentives (Tripathi, 2013), it has failed to do the needful and recently we have witnessed sporadic farmer protests in states like Punjab, Madhya Pradesh and Maharashtra, demanding loan waivers against crop loss and right price for their produce (Langa & Sriram, 2017; "Why the farmers stormed", 2018; Mohanty, 2018). On farmer distress, Swaminathan (2016, p.138), opined that "there is no other profession that has such low returns" and suggested that the government should set the MSP at the total cost plus 50 percent return.

There are many studies (Raghavan, 2008; Dev & Rao, 2010; Tripathi, 2013; Narayanamoorthy, Ali, & Suresh, 2014; Chand, Saxena, & Rana, 2015; Srivastava, Chand, & Singh, 2017) on costs and returns from crop cultivation. Almost all of these studies have found that the nominal cost of crop production has increased rapidly, while profitability of crop cultivation has decreased over time. However, these studies have mainly highlighted the national scenario.

In this chapter we would like to address the issues related to input cost and profitability in crop cultivation. As paddy is the main crop of our study area, we would like to give more importance on paddy cultivation. The main objectives of this chapter can be summarized as:

a) to assess the increase in the real cost as well as the nominal cost of paddy cultivation in the last twenty years, i.e., from 1996-97 to 2015-16;
b) to analyze the changes in returns from paddy cultivation during that time period;
c) to understand how farmers have been benefitted from MSP of paddy, and
d) to assess how farmers calculate cost of cultivation and analyze why they think that agriculture is still a profitable livelihood activity.

This chapter is mainly based on secondary data on the cost of cultivation (CoC)/cost of production (CoP) collected and published by the Directorate of Economics and Statistics, Ministry of Agriculture, Government of India. Time series data on CoC and CoP for paddy in West Bengal for the period of 1996-97 to 2015-16 is used here. We have been compelled to use state level data due to lack of block level or district level time series data on cost of cultivation. However, it is enough to get an idea about the profitability at the local level even from the analysis of state level data.

Data related to costs, returns and MSP have been deflated by the consumer price index for agricultural labourers (CPIAL) of West Bengal with the base year 2004-05 to examine how the costs, returns of paddy cultivation and MSP of paddy have changed in real terms. However, costs and returns at current prices/nominal prices[15] have also been presented. For analyzing CoC of paddy, the aforementioned time period (1996-97 to 2015-16) has been divided into two equal phases so that we can make a comparison. Besides, trend of MSP, ratio of MSP to cost of production, and ratio of realised price to MSP has been presented. In this connection it should be mentioned that MSP of paddy during *kharif* season has been taken into consideration.

The primary data has also been used to study the farmers' perception as well as to present cost calculation as done by them. In studying farmers' perception about profitability, the questions were set following 5-point Likert type scale. Two-way closed-ended questions (like yes/no) were also prepared on different cost items. A key informant survey was also conducted to gather information on how Assistant Directorate of Agriculture (ADA) staffs calculate costs and returns from crop cultivation.

Cost Items and Cost Concepts

Before discussing the profitability of paddy cultivation, it would be pertinent to know how many cost items are there and which types of cost concepts are used to assess profitability. There are two broad categories of cost items: i) paid out cost items, and ii) imputed cost items. Paid out cost items are those items for which farmers have to spend

15. In economics, current or nominal prices refer to the prices which have not been adjusted for inflation.

money or goods. These include hired labour, either human labour, animal labour or machine labour; maintenance expenses of own machine and animals; material inputs such as seed, fertilizers, manure, pesticides and irrigation; depreciation; rent paid for leased-in land; interest on working capital; and land revenue, cess and other taxes. On the other hand, imputed cost items are those items for which no money is actually spent but they do contribute towards the growth of crop. These items include family labour, interest on fixed capital, managerial input of the farmer, and so on. Proper evaluation of such items in terms of the monetary equivalent is vital for correct assessment of cost of production (Nath, 2008).

Based on these cost items, Commission on Agricultural Costs and Prices (CACP) has developed and used nine cost concepts. These are:

Cost A1	=	All actual expenses in cash and kind incurred in production by owner.
Cost A2	=	Cost A1 + rent for leased-in land.
Cost A2+ FL	=	Cost A2 + imputed value of family labour.
Cost B1	=	Cost A1 + interest on value of owned fixed capital assets (excluding land).
Cost B2	=	Cost B1 + rental value of owned land and rent for leased-in land.
Cost C1	=	Cost B1 + imputed value of family labour.
Cost C2	=	Cost B2 + imputed value of family labour.
Cost C2*	=	Cost C2 estimated by taking into account statutory minimum or actual wage whichever is higher.
Cost C3	=	Cost C2* + 10 percent of cost C2* on account of managerial input of the farmer.

However, the most used cost concepts are Cost A2, Cost C2 and Cost A2+FL, and thus have been used in this study. A2 cost is termed as paid out cost. C2 cost is more or less equivalent to the total cost. Cost A2+FL comprises A2 cost plus an imputed value of unpaid family labour.

Assessment of Cost of Paddy Cultivation in West Bengal

There are a number of studies on the profitability of Indian agriculture (Raghavan, 2008; Tripathi, 2013; Srivastava et al., 2017). From most of these studies, it has been observed that profitability of agriculture has declined considerably during the last few years, mainly due to increasing input cost leading to the agrarian crisis in India. Conversely, Chand et al. (2015) opined that there was no squeeze in the farmers' income or profitability in farming during 1983-84 to 2011-12; instead, per cultivator farm income increased by 7.3 percent a year.

We should now turn our attention to the cost of paddy cultivation. Narayanamoorthy (2013) in one of his study opined that though the area under paddy cultivation and production have increased from 1950-51 to 2008-09, the paddy farmers have either faced losses or realized a very marginal profit, especially after the 1990s. In the context of paddy cultivation in West Bengal, Srivastava et al. (2017) noted that in 2014-15, the cost of paddy production in this state was the highest among the 19 major paddy producing states in India. At the same time, the output cost ratio of paddy was lower than the output cost ratio of wheat, gram, *arhar* (*split toor* or *tuvar dal* known as pigeon peas), rapeseed and mustard, and sugarcane.

However, the trend in the average cost of paddy cultivation in terms of current prices and real prices in West Bengal during the past 20 years is shown in Fig. 4.1 (*see also*, Apendix 4.1). The trend in C2 cost at base year prices represents changes in real cost. It indicates that there is a marginal change in real cost of cultivation (CoC). During the period from 1996-97 to 2015-16, the nominal CoC of paddy has increased by 7.94 percent per annum, whereas the real CoC of paddy has increased only by 1.74 percent per annum (Table-4.1). This difference indicates that inflation of the inputs mainly has resulted in increased cost of paddy cultivation.

If we look into the changes in CoC in two time periods, it is observed (Table-4.1) that the recent time period has experienced a steeper rise in CoC of paddy. Compound annual growth rate (CAGR) in real C2 cost during 2006-07 to 2015-16 was 2.94 percent, which is three times greater than the CAGR in real C2 cost for the period of 1996-97 to 2005-06. Input use has experienced 2.5 times more growth during 2006-07 to 2015-16 as compared to 1996-97 to 2005-06.

However, the farmers are not concerned about the real cost. They are worried about the increase in nominal CoC. CAGR for A2 and C2 cost during the last period is more than double that of the previous period (*see*, part B of Table-4.1). Thus, it is evident that the cost of paddy cultivation has witnessed rapid growth during the last ten years caused by price rise as well as more input use.

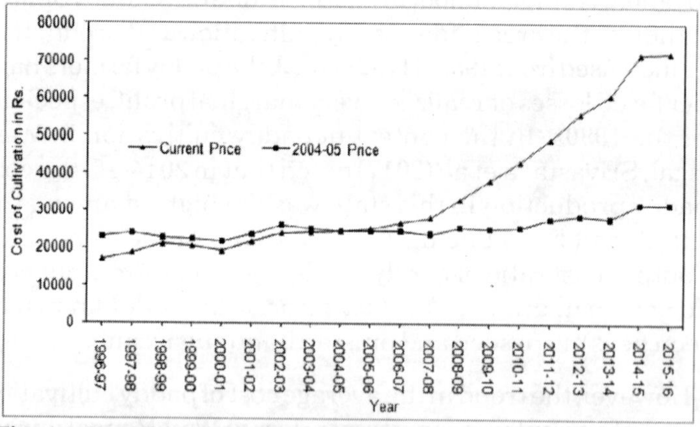

Fig. 4.1: Trend in C2 Cost of Paddy Cultivation in West Bengal (per hectare)

Source: Cost of cultivation data of various years published by Directorate of Economics and Statistics, Ministry of Agriculture and Farmers Welfare

Table-4.1
Average cost and CAGR in cost of paddy in West Bengal (Rs. per hectare)

Period	Average A2 cost	CAGR (%)	Average C2 cost	CAGR (%)
		A. Real Cost		
1996-97 to 2015-16	12787.2	2.15	25441.66	1.74
1996-97 to 2005-06	12168.8	1.31	23730.81	0.79
2006-07 to 2015-16	13405.5	3.32	27152.5	2.94

B. Nominal Cost				
1996-97 to 2015-16	18179.5	8.37	34714.52	7.94
1996-97 to 2005-06	11046.2	5.04	21512.2	4.50
2006-07 to 2015-16	25312.8	12.09	47916.84	11.68

Source: Computed from the cost of cultivation data of various years published by Directorate of Economics and Statistics, Ministry of Agriculture and Farmers Welfare.

Now, if we look into the different cost items (Table-4.2) it is observed that human labour is the main input as its share in total cost is highest. It is furthermore observed that the share of fertilizer, human labour, machine labour and cost of insecticides has increased over time, while share of manure, irrigation and bullock labour has declined. In general, advent of new technology helps to reduce use of manual labour. This is true for bullock labour, as the percentage share of cost of bullock labour has decreased with the increase of percentage share of machine labour. It appears from Table-4.2 that machines have failed to supersede the importance of human labour in paddy cultivation, as share of cost for human labour to C2 cost has witnessed more than 5 percent increase during 2006-07 to 2015-16 compared to the period of 1996-97 to 2005-06. However, this increase is a result of the increase in wages, as use of human labour in paddy cultivation in West Bengal has remained more or less constant for the whole time period. CAGR calculated at the current prices depicts that during 2006-07 to 2015-16, except for bullock labour other major input costs experienced more than two digits growth. For manure, cost share to total cost has declined but the average cost of manure for the period of 2006-07 to 2015-16 has increased compared to the previous decade. The primary reason for increase in input costs is the removal of input subsidies and the dependence of farmers on the open market for these inputs (Raghavan, 2008, p.125).

Profits from Paddy

"Returns from crop cultivation are essential not only for the survival of the farmers but also for facilitating reinvestment in agriculture" (Pushpa et al., 2017, p.388). From

Table-4.2
Growth in Cost of Different Inputs of Paddy Cultivation, West Bengal

Cost Items	Average Share to C2 Cost (%)		Average Cost of Inputs (Rs. per hectare)			CAGR (%)		
	1996-97 to 2005-06	2006-07 to 2015-16	1996-97 to 2015-16	1996-97 to 2005-06	2006-07 to 2015-16	1996-97 to 2015-16	1996-97 to 2005-06	2006-07 to 2015-16
Fertilizer	5.98	6.40	2198.9	1286.5	3111.4	8.26	2.4	12.38
Manure	2.31	1.96	699.9	486.8	913.0	5.46	2.98	10.61
Insecticides	1.01	1.30	442.0	216.3	667.8	13.1	8.13	18.94
Irrigation	4.84	4.03	1472.0	1039.8	1904.3	7.46	1.21	13.42
Bullock labour	9.45	5.14	2108.6	2071.9	2145.3	1.76	8.65	-3.23
Machine labour	2.31	4.23	1340.6	511.8	2169.5	15.8	10.3	21.66
Human labour	39.57	45.18	15303.3	8518.1	22088.6	9.39	5.4	13.57

Source: Computed from cost of cultivation data of various years published by Directorate of Economics and Statistics, Ministry of Agriculture and Farmers Welfare.

Table-4.3, it is observed that the return over paid out cost at current prices has increased gradually over time. However, it is actually an illusory increase as inflation is included there. Returns at real prices reveal the accurate picture that paddy cultivation was more profitable 15 years ago than at present. The pathetic condition of the paddy farmers is disclosed if we look into the return over C2 cost, either at current prices or at real prices. From 2000-01 to 2015-16, the farmers incurred losses in paddy cultivation upto 14 times except for in 2007-08 and in 2009-10. Average net returns over the whole time period are negative (Rs. -1577 per hectare). Once more, during the second half of the period under study, the farmers have suffered higher losses than during the first half, as during the first half average net return per hectare was about Rs. (-) 1100 while it was Rs. (-) 2053 per hectare for the second half. So, it can be concluded that paddy cultivation in West Bengal is unprofitable.

Table-4.3
Returns from Paddy Cultivation in West Bengal during 1996-97 to 2015-16

Year	Nominal Returns over paid out cost (Rs./Hectare)	Real Returns over paid out cost (Rs./ Hectare)	Nominal Returns over C2 cost (Rs./ Hectare)	Real Returns Over C2 cost (Rs./ Hectare)
1996-97	10788.20	14544.5	2159.68	2911.80
1997-98	10365.80	13327.5	1162.22	1494.24
1998-99	13327.10	14408.8	2890.92	3125.66
1999-00	10824.00	11895.7	363.66	399.67
2000-01	6672.73	7609.65	-2776.64	-3166.43
2001-02	7137.45	7844.13	-3077.74	-3382.50
2002-03	5736.83	6263.49	-5151.97	-5625.04
2003-04	8391.15	8704.84	-3024.43	-3137.38
2004-05	9823.63	9823.63	-1898.35	-1898.35
2005-06	10462.40	10187.00	-1778.15	-1731.40
2006-07	11468.40	10463.00	-1560.40	-1423.59
2007-08	14537.30	12255.50	512.78	432.29
2008-09	14782.90	11395.10	-425.01	-327.61

2009-10	21281.80	14061.20	3032.13	2003.39
2010-11	18872.50	11202.40	-1263.33	-749.88
2011-12	12925.00	7270.32	-8628.89	-4853.69
2012-13	19731.40	10031.40	-7084.59	-3601.72
2013-14	25489.40	11627.40	-3607.66	-1645.68
2014-15	21051.20	9284.84	-12884.8	-5682.88
2015-16	23302.70	10210.30	-10688.5	-4683.21

Source: Computed from the cost of cultivation data of various years published by Directorate of Economics and Statistics, Ministry of Agriculture and Farmers Welfare.

MSP, Realized Price and Profitability: Are Paddy Famers Deprived?

The government fixes MSP of major crops every year based on CACP's recommendation. Therefore, the trend of MSP and ratios of MSP to CoP would help us to understand whether MSP of the paddy has changed proportionately with the cost of production or not. The trend of MSP (Fig. 4.2) (Appendix 4.2) shows that nominal MSP of paddy has increased over time and it has experienced rapid increase after 2006-07. Conversely, real MSP has experienced a relatively lower rate of increase, with ups and downs. From

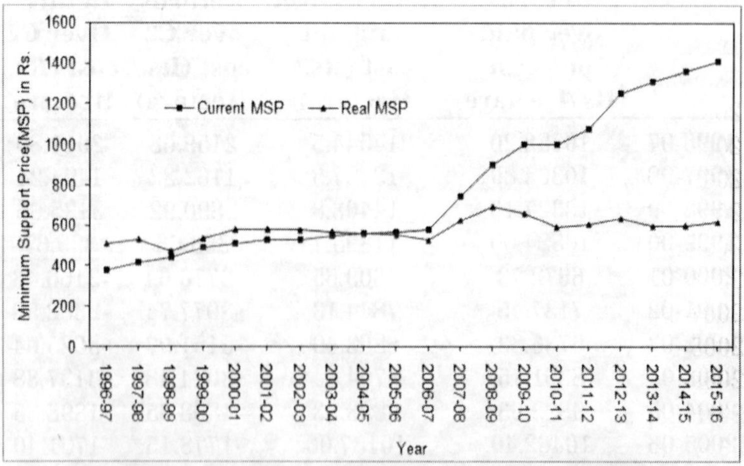

Fig. 4.2: Growth Trend of MSP of Paddy in West Bengal

Source: Directorate of Economics and Statistics, Ministry of Agriculture and Farmers Welfare

Table-4.4, it is observed that the ratio of MSP to C2 cost of production was either below 1 or close to 1 for the whole period, though the ratio of MSP to the A2 cost of production is well above 1.5. This indicates that paddy cultivation is an unprofitable activity if a farmer calculates C2 cost of production, but it is quite profitable if he calculates A2 cost of production, assuming MSP as selling price. However, CACP considers A2+FL cost of production to set the MSP of crops, and it is observed that ratio of MSP to A2+FL cost is more than 1 but less than 1.5 for most of the years, which means paddy cultivation fetched less than 50 percent return to the farmers.

Table-4.4
Changes in MSP of Paddy Relative to Cost of Production in West Bengal

Year	The ratio of MSP to C2 cost of production	The ratio of MSP to A2 cost of production	The ratio of MSP to A2+FL cost of production
1996-97	1.00	2.04	1.68
1997-98	0.97	1.90	1.54
1998-99	0.90	1.79	1.43
1999-00	1.00	2.04	1.54
2000-01	1.03	2.03	1.5
2001-02	1.06	2.01	1.49
2002-03	0.97	1.79	1.35
2003-04	1.01	1.93	1.43
2004-05	0.97	1.90	1.42
2005-06	0.98	1.92	1.44
2006-07	0.93	1.82	1.36
2007-08	1.11	2.26	1.67
2008-09	1.23	2.28	1.74
2009-10	1.16	2.21	1.65
2010-11	0.98	1.83	1.34
2011-12	0.99	1.78	1.34
2012-13	1.01	1.94	1.38
2013-14	0.96	1.84	1.33
2014-15	0.96	1.84	1.28

2015-16	0.99	1.88	1.33

Sources: 1. Computed from the cost of cultivation data of various years published by Directorate of Economics and Statistics, Ministry of Agriculture and Farmers Welfare.
2. Ministry of Agriculture, Government of India.

However, the farmers are more worried about the price they realize rather than the MSP (Dev & Rao, 2010, p.177). Therefore, the ratios of prices realized by the farmers to MSP of the paddy have been presented in Table-4.5. The ratio of less than 1, indicates realized price less than MSP. This table reveals that realized price was higher than the MSP only for five years out of 20 years during 1996-97 to 2015-16. A similar situation is observed in Orissa, Assam, Bihar and UP, whereas paddy farmers received more than MSP in states like Haryana, Himachal Pradesh, Punjab, Madhya Pradesh, Chhattisgarh, Karnataka, Kerala, Andhra Pradesh and Tamil Nadu (Dev & Rao, 2010, p.177). One probable cause of such realised prices is that during the harvesting period, the local market price of paddy remains low; big traders further suppress the paddy price. Thus, resource-poor small and marginal farmers, having cash requirement, are compelled to sell their produce at prices less than MSP. In this way, the paddy growers of West Bengal are deprived by the government as well as by the local traders.

Table-4.5

Changes in Realized Prices in relation to MSP of Paddy in West Bengal

Year	Price realized at current price (Rs. per quintal)	The ratio of realized price to MSP
1996-97	438.77	1.15
1997-98	461.93	1.11
1998-99	572.74	1.30
1999-00	500.77	1.02
2000-01	409.05	0.80
2001-02	415.27	0.78
2002-03	411.19	0.78

2003-04	465.86	0.85
2004-05	525.74	0.94
2005-06	533.63	0.94
2006-07	583.29	1.01
2007-08	682.35	0.92
2008-09	720.38	0.80
2009-10	943.94	0.94
2010-11	988.70	0.99
2011-12	870.88	0.81
2012-13	1060.02	0.85
2013-14	1275.68	0.97
2014-15	1130.26	0.83
2015-16	1190.27	0.84

Source: Computed from the cost of cultivation data of various years published by Directorate of Economics and Statistics, Ministry of Agriculture and Farmers Welfare.

Farmers' Perceptions on Profitability of Paddy Cultivation

From the above discussion, it is very clear that paddy cultivation is now a loss-making economic activity. Thus, the question arises, why are farmers sticking to paddy cultivation? In any way, is paddy cultivation a profitable venture for the farmers? In response to these questions, the farmers responded differently to the same question (Table-4.6).

Interestingly, only 7.8 percent farmer households opined that paddy cultivation incurred loss. About 37.2 percent farmer households opined there is either no profit or no loss in paddy cultivation. About 46.3 percent farmer households perceived that paddy cultivation fetched very low to low-profit margin to them. The causes behind such perceptions are not same for all types of farmer households. Landless lessee and sharecropper households have blamed that landowners leased out relatively bad landholdings to them. Besides, they have to pay share of output to the landowner. All these factors lessen their profit. However, according to other categories of farmer households, full dependence on hired labour and hired farm implements are most crucial causes for paddy cultivation an unprofitable venture.

Table-4.6
Farmers' Perceptions about Profitability of Paddy

Farmer's Perceptions	Farmer Household Categories (%)						
	Landless lessee	Landless Sharecropper	Marginal	Small	Medium	Big	All
Loss	0	10.0	6.4	11.8	7.1	10.5	7.8
No loss no profit	33.3	45.0	36.7	42.1	31.0	31.6	37.2
Very low profit margin	26.7	25.0	34.6	23.7	21.4	26.3	29.4
Low profit margin	20.0	15.0	15.4	13.2	33.3	10.5	16.9
Moderate to high profit margin	20	5.0	6.9	9.2	7.1	21.1	8.6

Source: Field Survey, 2016

Only a few farmer households (8.5 percent) believed that paddy cultivation is moderate to highly profitable. About 21.1 percent of the big farmer households and about 20 percent of the landless lessee farmer households too perceive so, though they do not provide reasonable arguments in favour of high profitability of paddy cultivation. They mentioned that productivity has increased in comparison to 20-25 years ago, but as consumption expenditure per households has increased rapidly over time, it apparently seems that profitability of paddy cultivation is going down.

Many researchers have proved that profitability of a crop may vary among the different categories of farmers depending on labour use, input use, technology use (Gaurav & Mishra, 2011; Chand et al., 2011; Ranganathan, 2015). Besides, from the farmers' cost analysis (Table-4.7), it has been observed that profitability varies based on seed varieties and crop seasons. However, the important question is: why does the perception of the farmers differ from experts' opinion to such a large extent?

The answer to this question is perhaps that except for a few farmers, the others miscalculate the cost of cultivation. The leading cause of such exaggerated profit is that they only include the input costs of seed, fertilizers and pesticides, irrigation, hired human and machine labour (Table-4.7). Besides, farmers do not keep detailed records of all costs incurred, and they calculate the cost arbitrarily. They do not include cost items like family labour, own farm yard manure or own machine use, the maintenance cost of the farm implements. Most of the farmers do not know what depreciation cost is or what rent on fixed capital is. Interestingly, the majority of them do not even include the cost of bags used for packing the grains. Exclusion of these cost items lessens the apparent cost of cultivation.

Table-4.7

Cost Calculation of Paddy (per *bigha) by farmers**

For *Boro* paddy cultivation
(If they cultivate minikit/IET 4786 variety)

Input	Output
Ploughing = Rs. 500 Fertilizer = Rs. 700	13 packets of paddy/*bigha*
Pesticides = Rs. 100	Value = 13 packets × Rs. 900 = Rs. 11,700
Labour (30 man days) = Rs. 4,500	Straw = 2 *kahon***
Seed and Irrigation = Rs. 1,000	Value = 2 *kahon* × Rs. 300 = Rs. 600
Total = Rs. 6,800	Total = Rs. 12,300
Profit = Rs. 5,500/*bigha* (Rs. 34,375/hectare)	

For *Aman* paddy cultivation
(If they cultivate swarna variety)

Input	Output
Ploughing = Rs. 600 Fertilizer = Rs. 350	13 packets of paddy/*bigha*
Pesticides = Rs. 300	Value = 13 packets × Rs. 780 = Rs. 10,140
Labour (25 man days) = Rs. 3,750	Straw = 2.5 *kahon*
Seed and Irrigation = Rs. 840	Value = 2.5 *kahon* × Rs. 300 = Rs. 750
Total = Rs. 5,840	Total = Rs. 10,890
Profit=Rs 5,050/ bigha (Rs. 31,562/hectare)	

* 1 *Bigha* = 0.1338 hectare. However, in this calculation 1 bigha is considered as equal to 0.16 hectare because farmers of the study area normally follow this calculation.

**1 *Kahon* = 1280 piece of straw bundles; it is more or less 320 kg in weight.

Source: Author's calculation based on farmers' interviews, May and December 2016.

Another reason for overstated profit is that farmers incorporate income from paddy business as return from agriculture. According to the ADA officials, if a farmer stores his produce at home for a few months and sells it after the rise of market prices, then it should be considered as income or profit from business. This profit should not be included as return from cultivation.

Profit from Mustard Cultivation and Farmers' Perceptions

Returns from the mustard cultivation in West Bengal (Table-4.8) reveal that mustard cultivation is profitable for all the years if paid out cost, either in nominal term or in real term, is considered, and profitability varies year by year with ups and downs. However, real returns over C2 cost disclose that out of 17 consecutive years, farmers incurred losses 5 times, and mustard cultivation is also not so profitable. Moreover, fluctuating trend of returns indicates mustard cultivation also do not provide any assured profit to the farmers.

Table-4.8
Returns from Mustard Cultivation in West Bengal during 1999-2000 to 2015-16

Year	Nominal Returns over paid out cost (Rs./Hectare)	Nominal Returns over C2 cost (Rs./ Hectare)	Real Returns over paid out cost (Rs./ Hectare)	Real Returns Over C2 cost (Rs./ Hectare)
1999-2000	3928.96	-1395.16	4318.01	-1533.31
2000-01	6385.79	33.04	7282.23	37.68
2001-02	4190.29	-1685.45	4605.22	-1852.35
2002-03	8540.25	876.87	9324.43	957.39
2003-04	12328.16	3310.75	12788.55	3434.39
2004-05	11528.2	3084.86	11528.20	3084.86
2005-06	7444.16	-335.96	7248.45	-327.13
2006-07	6047.9	-1932.05	5517.65	-1762.66
2007-08	13516.77	2439.31	11395.02	2056.41
2008-09	9734	-1168.41	7503.28	-900.65

2009-10	14456.25	2650.69	9551.54	1751.36
2010-11	21369.78	4330.46	12684.62	2570.46
2011-12	19069.17	3227.54	10726.27	1815.47
2012-13	21791.84	4428.16	11078.72	2251.23
2013-14	19479.84	1422.12	8885.98	648.72
2014-15	23713.28	3219.21	10458.82	1419.84
2015-16	24334.95	1702.5	10662.47	745.96

Source: Computed from the cost of cultivation data of various years published by Directorate of Economics and Statistics, Ministry of Agriculture and Farmers Welfare.

As per field data, about 33.3 percent farmer households used to cultivate mustard. Percentage of farmer household cultivating mustard has remained unchanged over last 20-25 years from the date of survey. However, more farmer households (44 percent) in Galsi-II CD Block used to produce mustard compared to the farmer households (23.2 percent) in Galsi-I CD Block. Furthermore, percentage of farmer households cultivating mustard in Galsi-I CD Block has decreased over time, whereas it shows an increasing trend in Galsi-II CD Block. Regarding the change of mustard cultivation, farmers opined ratio of area under mustard to potato per farmer household has decreased. Over the years potato cultivation has increased in place of mustard as the former incur more profit in spite of huge market risk involved. Besides, production of mustard declined if same land is used for mustard cultivation year after year, but there is no such problem in potato cultivation. Disease related to swelling root of mustard plant is very common in our study area which resulted in low production, thus low returns. Low cost palm oil, decreasing use of mustard oil caused by change in oil consumption etc. are also indirectly responsible for decline of area under mustard cultivation.

Profits from Potato

Potato is one kind of important cash crop of West Bengal. In the study area, potato cultivation experiences growth in terms of area as well as production. In many parts, potato cultivation has replaced *boro* paddy cultivation. Returns from potato cultivation in West Bengal during 1996-97 to 2015-16 (Table-4.9) reveal that this crop procures more returns to the farmers compared to returns from paddy or mustard over paid out cost. However, out of 20 years, farmers incurred loss in 2 years over A2 cost and faced loss 13 times out of 20 years if returns over C2 cost are considered.

Table-4.9
Returns from Potato Cultivation in West Bengal during 1996-97 to 2015-16

Year	CPI_AL base 2004-05	Nominal returns over paid out cost (Rs./ Hectare)	Nominal returns over C2 cost (Rs./ Hectare)	Real returns over paid out cost (Rs./ Hectare)	Real returns over C2 cost (Rs./ Hectare)
1996-97	74.17	9966.38	-6229.47	13437.21	-8398.91
1997-98	77.78	15419.34	-1262.26	19824.30	-1622.86
1998-99	92.49	8751.09	-9989.43	9461.66	-10800.55
1999-2k	90.99	4759.45	-10974.17	5230.74	-12060.85
2000-01	87.69	15254.99	-1978.46	17396.50	-2256.20
2001-02	90.99	29923.24	8388.8	32886.30	9219.47
2002-03	91.59	1643.65	-13334.4	1794.57	-14558.79
2003-04	96.4	33397.21	10318.17	34644.41	10703.50
2004-05	100	17660.03	-1712.77	17660.03	-1712.77
2005-06	102.7	41025.87	12973.48	39947.29	12632.41
2006-07	109.61	13953.52	-8455.31	12730.15	-7714.00
2007-08	118.62	4106.37	-17301.18	3461.79	-14585.38
2008-09	129.73	21186.88	-2291.71	16331.52	-1766.52
2009-10	151.35	-5923.32	-27083.65	-3913.66	-17894.71
2010-11	168.47	34750.17	2918.55	20626.92	1732.39
2011-12	177.78	53916.86	12510.6	30327.85	7037.12
2012-13	196.7	27700.47	-13751.3	14082.60	-6991.00
2013-14	219.22	47734.64	887.4	21774.77	404.80
2014-15	226.73	-7927.18	-47193.78	-3496.31	-20814.97
2015-16	228.23	98750.38	36030	43267.92	15786.71

Source: Computed from the cost of cultivation data of various years published by Directorate of Economics and Statistics, Ministry of Agriculture and Farmers Welfare.

Conclusion

It can, therefore, be concluded that farmers make many mistakes in their cost calculation as they do not count their own input cost that results in reporting of exaggerated profits from paddy cultivation. This type of faulty calculation also helps them to continue paddy cultivation at the time when experts opine it as unprofitable. However, most of the farmers perceive that they are continuing paddy cultivation as it helps to meet the family needs either partly or wholly, and there is a lack of opportunity for engaging in other gainful economic activity due to various causes.

The cash crops like potato and onion are gaining importance in the study area like in West Bengal, but these crops have also failed to fetch considerable profits regularly in recent times. Moreover, shifting from food crops to cash crops may create food insecurity in future. One-time incentives like loan waivers is also not a solution, as most of the marginal and small farmers either do not get institutional loans or prefer to repay the loans before/in time anyhow.

There are many problems in paddy procurement scheme of the government. Most of the marginal and small farmers have alleged corruption in the distribution of coupons required for selling paddy at procurement centres. Further, there is presence of middlemen in the procurement system. The government has also failed to procure paddy immediately after harvesting. Besides, farmers blamed that procurement agencies are levying an unlawful cut on amount of paddy. Farmers are also forced to bear high transportation cost to bring their produce to procurement centres.

However, the state as well as central government, have recently announced that they would provide a sum of money every year as direct incentives to farmers ("Krishak Bandhu", 2018; "Budget 2019", 2019). This is also good for the farmers, but if the government would take steps to provide the right price of produce (at least 50% profit over C2 cost) then profit would be far higher than the number of incentives. Finally, the government should train and empower farmers so that they can adequately calculate the cost of cultivation because it is essential to assess the cost and returns from an economic activity before engaging in it.

The next chapter will discuss the government support mechanisms to the farmers with a special focus on the limitations of those schemes. It will help us to perceive how majority of the farmers are being deprived from the facilities provided under the various government schemes and by the different institutions.

5 Government Support Mechanisms

Introduction

Like other agriculturally developed countries of the world, India also supports agriculture and the farmers either to ensure food security and/or to enhance the income levels of the farmers. The government helps the farmers in two ways. One is through providing input subsidies, and the other is through providing minimum support prices for the major crops. Input subsidies include fertilizer subsidy, irrigation subsidy, credit subsidy, etc. Besides, the government also invests in improving infrastructure for agricultural development. The government provides funds for all subsidies and infrastructure development through different schemes. Different government institutions like Block Development Office, Assistant Directorate Office, Co-operative and banks, Panchayats, played an important role in village level implementation of such programmes and various policies related to agriculture. However, in an analytical study related to subsidy on agriculture, Salunkhe and Deshmush (2014) opined due to corruption and ineffective management of subsidies in India, and due to the illiteracy of farmers regarding agriculture subsidies, it has not reached to the farmers. The question of subsidies in farming has appeared as an important issue in ongoing policy debates (Dev, 2008). From the various studies, it is observed that subsidies and other government supports do not reach evenly to the farmers of all regions and of all categories.

In this context, an attempt has been made to assess the

government supports in our study area. In the previous chapter we have discussed the profitability of different crops, and to do this we have also discussed about the minimum support price (MSP). Therefore, in this chapter, we would like to give insights into some of the schemes related to agricultural development. Besides, we would like to highlight the role of primary agricultural credit societies (PACS) in agricultural development, and issues related to the office of Assistant Director of Agriculture (ADA) and Krishi Prayukti Sahayak (KPS).

This chapter is mainly based on primary data obtained from field surveys and key informant surveys. However, in some cases, state level and/or block level secondary data has also been used. Block level official data, either due to non-availability or due to lack of reliability, has been used minimally.

Agricultural Credit and Indebtedness

Like any other system, credit is the most important input in agriculture. As in India, majority of the farmers are small and marginal, credit flow is the primary concern as they need external finance to enable cultivation. There are two segments of credit structure in India: 1) Informal Segment and 2) Formal Segment. The informal sector comprises people and firms that are not enlisted with the government as financial intermediaries and are not subject to government supervision. The main characteristics of this sector are a) easy to enter and exit b) informal transaction c) small scale transaction etc. Nowadays, rich farmers, *aratdars*, friends, relatives, rural businessmen etc., formed the dominant entity of the informal segment. Historical evidences tell us (see chapter 2) that Indian farmers have long been exploited by the money lenders, rich farmers, landlords etc., due to insufficient cash in hand. The Government of India, as a result, has tried to provide credit support to the farmers. This type of credit is known as institutional credit or formal sector credit source. The formal segment works within the provision of the Indian Banking Companies Act (1956) and keeps up accounts that are exposed to audit and regular assessment (Singh & Asokan, 1988). Generally, banks, co-operative societies, self-help groups, etc., are the sources of formal loans to rural people. Availability and access to adequate, timely and low-cost credit from formal sectors are of significance particularly to small and marginal farmers (Ramakrishna, 2016).

In the study, it has been observed that about 59 percent of the farmer households have taken a loan. The main source of loan is the bank. Other important sources of formal sector loan to the farmer households are co-operatives and self help groups. About 10.3 percent and 4.7 percent farmer households have taken a loan from co-operatives and self-help groups respectively. Thus, about 45 percent farmer households have access to the formal sector loan. Only 10.8 percent farmer households depend on *aratdar* or *mahajan* for loan (Table 5.1). Interestingly, it is the second most important source of loans to farmer households.

Moreover, there is a difference among farmer households in terms of source of loan. No one among the medium and big farmer households has taken a loan from *aratdar* or *mahajan*. Farmer household categories who mostly depend on *aratdar/mahajan* for a loan are landless lessee followed by the marginal farmer and landless sharecropper. These three categories of farmer households have the least access to the bank or co-operatives for loan. It is the big farmers who have maximum access to the institutional credit source. Among big farmer households, about 42 percent depend on the bank, and 26.3 percent depend on co-operatives for the agricultural loans. Rich and educated farmer households have easy access to formal sector loans as belonging to hegemonic class bank officials and co-operatives favour them. Less-educated resource-poor farmers remain in the same position as they were. In this connection, it should be mentioned that many of the middle-class farmer households are not interested to take any loan, rather they prefer to manage the credit required for farming anyhow without taking a loan from banks or *aratdar*. They do not want to be involved in any kind of risk

Table-5.1

Distribution of farmer households (%) by the source of the loan

Sources of Loan	Landless lessee	Landless Sharecropper	Marginal	Small	Medium	Big	All
Bank	26.67	20	27.13	36.84	30.95	42.11	30.00
Co-operatives	6.67	0	7.45	18.42	7.14	26.32	10.28
SHG	0	0	5.32	3.95	7.14	5.26	4.72
Relatives/trader	0	30	2.66	0	0	0	3.06
Aratdar/Mahajan	20.00	10	14.36	9.21	0	0	10.83
No loan taken	46.67	40	43.09	31.58	54.76	26.32	41.11

Source: Field Survey, 2016-2018.

or in any sort of financial transaction.

Many farmer households prefer informal sources of loan over formal sector loans as they think taking a loan from a bank is harder than the informal sector. About 48 percent farmer households (Fig. 5.1) opined that it is hard to get formal sector loans. The majority of the landless sharecroppers, landless lessee farmers and marginal farmers, think getting a formal sector loan is not accessible. Besides, informal sectors take less time to provide loan and require few or no documents and less paperwork. Moreover, relatives, friends or sometimes traders offer a loan without any interest for having good personal relationships. The amount of loan provided to the farmer by the formal sectors is also a concern. One marginal farmer of Ramgopalpur village informed that he was compelled to take a loan from *mahajan* along with the formal sector, because the bank provides crop loans based on the land amount which seems insufficient to him as he needed money to meet the family expenditure also until harvesting of paddy.

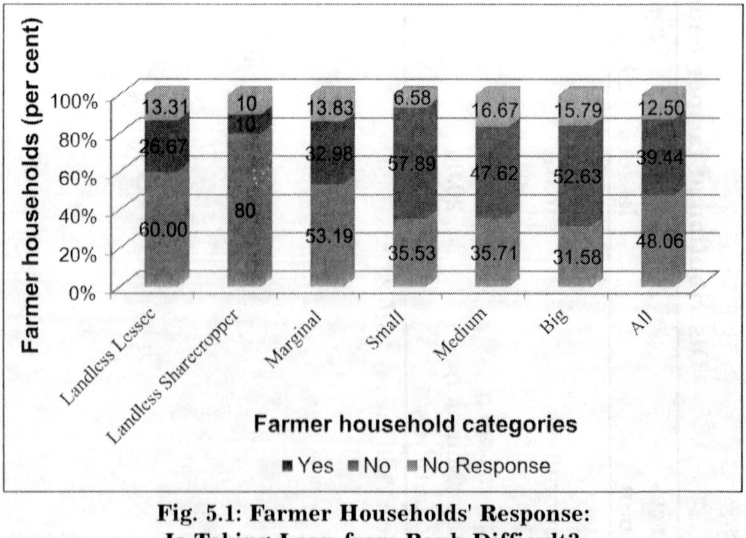

Fig. 5.1: Farmer Households' Response: Is Taking Loan from Bank Difficult?

Source: Field Survey, 2016-2018.

If we look at the interest rate of the loan, we will see that as per expectation, the interest of formal sector loans is very low as compared to the informal sector loan. Formal sectors provide short term crop loans up to 3 lakh at 7 percent rate of interest. Besides, there is interest subvention scheme. Under this scheme, the farmer has to pay only 4 percent interest per annum. However, interest rates may vary from one commercial bank to another, from one formal credit source to another. It has been reported that many banks, co-operatives and SHGs fixed interest rates at more than 7 percent. On the other hand, the interest rate of the loan taken from *mahajan* varies from 2 percent to 5 percent per month, i.e. 24-60 percent per annum. In this case, interest rates also differ from one man to other. *Mahajan* charges less interest rates to the known persons who have a good track record in loan repayment. It has also been observed that *aratdars* provide a loan without any interest on the condition that farmers should sell off their produce to them.

Kisan Credit Card

"Government of India introduced the Kisan Credit Card scheme (KCC) in 1998 as an innovative credit delivery mechanism to enable the farmers to meet their production credit requirements in a timely and hassle-free manner" (Mani, 2016, p.1). The primary purpose of this announcement was to discourage distress sales. At present, a farmer possessing KCC can get a loan for short term credit requirements for crop cultivation, post-harvest expenses, marketing of crop produced, household consumption requirement, investment credit requirement in agriculture and allied activities, etc. Besides, a farmer can insure his crop and his life, his health, even his assets with the help of KCC. All farmers who own land are eligible for this scheme. Tenant farmers, oral lessees, or sharecroppers also can get KCC individually or through an SHG or joint liability group (JLG). One farmer may have two KCC. Mani (2016) observed that 'some of the farmers had taken one from co-operatives and the other from either a commercial bank or a regional rural bank'. KCC is valid for five years and subject to an annual review. Crop insurance is mandatory for KCC holders while accident insurance, health insurance or asset insurance are optional. Therefore, it is clear that if a farmer possesses KCC, he has ample scope for achieving financial security for his family. Applying for a KCC needs very few documents such as identity proof, a passport size photo, address proof, land documents are required. A declaration and written consent of the landowner on registry paper is required for farmers who do not have land documents in their own name.

Realizing the importance of KCC, we have collected information about it and presented in Table-5.2. From this table, it has been observed that only 39.72 percent farmer households possess KCC. The distribution of KCC is uneven among the different farmer household categories. Though there is a provision of getting KCC for all, no one among the surveyed landless lessees possesses this. Only 10 percent of the landless sharecroppers and 33.5 percent marginal farmer households have KCC. Most of the big farmer households (73.68 percent) possess KCC. Therefore, it can be said that the landless lessees, the sharecroppers and even most of the marginal farmers have failed to access government supports provided through KCC.

Table-5.2

Response of Farmer Households on KCC (%)

Farmer household categories	Have KCC	Don't have, wish for it but do not applied yet	Don't have but do not know how to get KCC	Don't want to get KCC
Landless Lessee	0	0	6.7	93.3
Landless Sharecropper	10	20	60	10
Marginal	33.51	12.77	47.34	6.38
Small	56.58	22.37	15.79	5.26
Medium	50.00	16.67	30.95	2.38
Big	0	73.68	15.79	10.53
All	39.72	14.72	39.72	5.83

Source: Field Survey, 2016-2018.

About 14.7 percent farmer households know how to apply for KCC but do not apply for it. The leading cause of such behavior is that most of them do not have land-related documents in their name. Besides, some of the farmers opined that applying for a KCC is just a loss of time because bank officials will reject their application anyhow. About 39.72 percent farmer households do not know how to apply for KCC and what documents are required for it. About 93.3 percent of the landless lessees, 60 percent of the sharecroppers, and 47 percent marginal farmers do not know how to apply and to whom for obtaining a KCC. It hints that there is a lack of administrative initiatives in creating awareness among farmers about KCC. However, it is very interesting to note that about 5.83 percent farmer households responded that they do not want KCC, because they do not wish to fall into the trap of loan. Most of them have a perception that taking a loan is not a good thing.

The farmers who possess KCC also have a lack of knowledge regarding the facilities they can avail of with this card. Though farmers used to take a loan using KCC, many of them do not know that crop insurance is mandatory and they pay for it. In many cases, bankers debit the premium for accident insurance scheme, weather-based crop insurance scheme without the knowledge of farmers. Many farmers are reluctant to get these schemes as they rarely avail compensation from these schemes. Farmers also claimed that they do not enjoy the facility provided by the interest subvention scheme. As the bank offers a very less period of repayment for enjoying interest subvention, most of the farmers do not get it. However, recently Reserve Bank of India (2019) said that small and marginal farmers having KCC would get the interest subvention of 3 percent for a period of up to six months after the harvest of the crop.

Crop Insurance

In India, crop insurance is one of the instruments for defending farmers from agricultural inconsistency (Sinha, 2004, p.2605). In 1985-86, General Insurance Corporation first introduced a crop insurance scheme in India, named Comprehensive Crop Insurance Scheme. In the year 1999-2000, this scheme was replaced by the National Agricultural Insurance Scheme (NAIS). Other existing crop insurance schemes are the Modified National Agricultural Insurance Scheme (MNAIS), Weather Based Crop Insurance Scheme (WBCIS), Coconut Palm Insurance Scheme (CPIS) etc. However, NAIS failed to achieve considerable success regarding crop insurance coverage. Sinha (2004) pointed out that NAIS covered roughly 10 percent of the gross cropped area in 2002. According to Sinha, implementing agencies neither bear any risk nor earn any returns from this scheme, and no ways to expand crop insurance coverage. Rural financial institutions, i.e., co-operative credit societies, the rural branches of commercial banks, and regional rural banks, were responsible for the disbursement of the loan. These rural financial institutions, especially co-operative credit societies, failed to function properly. The major faults of the NAIS were the goal of economic feasibility, its obligatory nature, its failure to tackle adverse selection, the area approach, and arbitrary premiums. "Like other agriculture related schemes ranging from fertilizer subsidies to subsidised loans and loan waiver for farmers, crop insurance schemes have witnessed that the benefits are highly skewed in favour of just few states and only the large and wealthy farmers." (ICFA report on Crop Insurance in India, 2016, p.10).

A new scheme, namely, Pradhan Mantri Fasal Bima Yojana (PMFBY) has been implemented, replacing MNAIS/NAIS from *Kharif* 2016 season. Under PMFBY, farmers have to pay only 2 percent premium for all *kharif* crops, 5 percent for annual commercial and horticultural crops, and 1.5 percent for all *rabi* crops. This insurance scheme is mandatory for the farmers who possess a crop loan or KCC account (called loanee farmers) and optional for the other farmers (non-loanee farmer). However, The PMFBY is also criticized because of delays in settlement of claims from farmers. "Farmers in some places are not receiving compensation from insurance firms due to delayed contributions by state governments and consequent claim rejections by insurers" (Mishra, 2019, April 1 para7).

In West Bengal, Pradhan Mantri Fasal Bima Yojana (PMFBY) is known as the Bangla Fasal Bima Yojana (BFBY) as the government pays the farmers' share of premium except for sugarcane and potato. For these two crops, farmers' share of the premium is 8.85 percent. "The West Bengal government has stopped implementing the Central scheme on crop insurance PMFBY from *Kharif* 2019 and instead rolled out its own scheme named Bangla Shasya Bima Scheme" (Press Trust of India, 2019, July 23). In the 1st phase, the scheme would remain applicable in 15 districts and the number of districts to be increased in the next (2nd) phase. Farmers will not have to pay any premium to avail the Bangla Shasya Bima Scheme benefits. "On behalf of the farmers, the state govt. will bear the Bangla Shasya Bima crop insurance premium" (https://sarkariyojana.com). Insurance agency for Jalpaiguri, Alipurduar, Uttar Dinajpur, Bankura, Howrah, Jhargram and Paschim Medinipur is the Oriental Insurance Company Limited. Agricultural Insurance Company of India Limited is the insurance agency for the rest of the districts of West Bengal. *Aman* paddy, *aus* paddy, jute and maize are the crops under this scheme. For monitoring and supervision of the scheme there are state level, district level and block level committies. In the Block level monitoring committee, Block Development Officer is the chairman and Assistant Director of Agriculture (ADA) is the convener. Other members in this committee are Krishi Karmadhyksha, Sech-o-Samabay Sthayee Samiti of the respective Panchayat Samity, Co-operative Inspector, representative of implementing insurance agency, representative of local reputed NGO involved with farmers' welfare, and one farmer representative. This block level monitoring committee has to prepare the list of cases of crop loss. In the case of localized crop failures and post-harvest calamities, preliminary information has to be furnished by the farmer to the insurance company, bank, or ADA within 72 hours of the occurrence of natural hazards.

Roy et al. (2018) showed that more than 70 percent of the non-insured farmers have not even heard the name of PMFBY or BFBY, and most of the farmers are not aware of the features of this crop insurance scheme. However, in our survey, we did not assess the awareness of the farmers about crop insurance, but asked if they had received compensation for crop loss. It has been observed that as crop insurance is mandatory for the loanee farmer, their crops were insured, and non-loanee farmers did not insure their crops separately. Farmer of both of these categories opined that crop insurance is nothing but an extra cost. Many marginal and small farmers told that they have a lack of money even to insure their lives and how they can insure their crops. Many loanee farmers under marginal and small farmer categories opined that getting compensation for crop loss is not easy if crop loss occurred due to localized calamities. During our first survey, MNAIS and WBCIS were operating in our study area. About 39.7 percent of the loanee farmers were under insurance cover automatically. However, these crop insurance schemes were not popular among farmers. Following Gulati et al. (2018) it can be said that the leading cause for low coverage under NAIS was higher rates of premium and capping of premium and sum assured under MNAIS.

After the implementation of PMFBY/BFBY/BSB, coverage of farmers under this scheme has increased within a short time as farmers do not have to pay any premium for this. However, delay in claim settlement has been reported due to the state government's delay in payment of their share of premium and delay in crop cutting experiments to estimate yield loss.

Loan Waiver Scheme

Loan waiver is another instrument for tackling agricultural distress. The central government or state government may implement loan waiver scheme. In 2008, Agricultural Debt Waiver and Debt Relief Scheme was implemented under the UPA government. It was the first large scale debt relief programme after Agricultural and Rural

Debt Relief Scheme of 1990 (Ramakumar, 2017). There were logics for the implementation of this scheme. One of the rationales was the falling rate of profitability due to an unparalleled increase in input cost compared with the output and market price. However, Ramakumar (2017) opined that West Bengal ranked first in terms of the number of beneficiaries who received a full waiver under that loan waiver scheme. About 98.9 percent of the indebted farmer households were granted a full exemption in West Bengal. Most of these beneficiaries were small and marginal farmers. Indebted farmers who took loan from informal sources did not enjoy this facility.

In this research, it had been tried to find out who enjoyed the loan waiver under this scheme. It has been observed that 45 farmer households out of 360 farmer households received loan waiver. It accounts for only 12.5 percent of the total farmer households surveyed. Such low percent of beneficiaries to total farmer households are caused by three factors. Firstly, indebtedness among the farmer households in West Bengal is not very high. As per Situation Assessment Survey of 2003, only about 50 percent of the farmer households in West Bengal were indebted. Secondly, among the indebted farmer households, more than half of them depend on informal sources of loan. About 58 percent of the indebted farmer households in West Bengal accessed formal sources of loan (NSSO 59th Round, 2003). Thirdly, as it is mentioned before, poor farmers try to repay the loan amount anyhow before the stipulated time in spite of poor financial conditions.

Out of 18 villages surveyed, farmer households of 12 villages received loan waiver. However, the distribution of farmer households is very uneven. Among these 12 villages, Sanota ranks first in the number of farmer households received loan waiver, and in rest of the villages less than 25 percent of the surveyed farmer households benefitted from this scheme. At Sanota village, 14 out of 20 farmer households enjoyed this facility.

However, among the different farmer household categories, no one of the landless lessees and landless sharecroppers receives loan waiver (Fig. 5.2). The ratio of beneficiary farmer households to total farmer households is highest for big farmer households. About 26.3 percent of the big farmer households received a full loan waiver. Proportions of marginal, small and medium farmer households to corresponding total households are also negligible. Having better access to formal sector loan, most of the big farmers have enjoyed a full exemption under the loan waiver scheme.

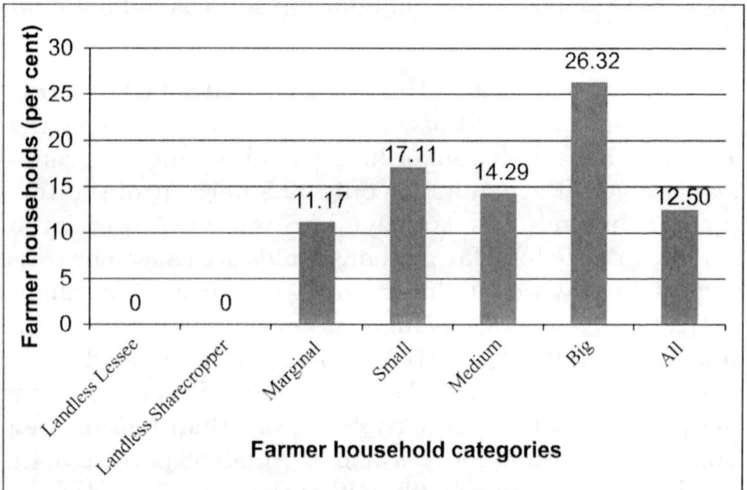

Fig. 5.2: Category-wise Distribution of Farmer Enjoyed Loan Waiver in the Study Area

Source: Field Survey, 2016-2018

Special Compensation for Crop Loss in the Natural Calamity

During 2012-16, the Government of West Bengal had provided Rs. 1336.43 crore as agricultural input subsidy among 32.83 lakh farmers for crop loss caused by various natural calamities (Bengal Government to Pay Compensation, 2017). In 2015, the West Bengal government announced compensation to farmers who had lost more than 33 percent of their crops due to flood under the State Disaster Response Fund. As our study area was affected by heavy rain and flood, farmers applied for compensation. Therefore, we enquired about it. From the study, it is observed that only 27.22 percent farmer households received compensation. More than 30 percent of the marginal and big farmer households got compensation. However, only 6.7 percent of the landless lessee households and 10 percent of the sharecropper households benefitted from this facility. It is exciting to note that percent of affected farmer households was much higher than the farmer households received compensation and were fit for availing compensation under this scheme.

From interviews, we came to know that many farmer households failed to apply for compensation because of the lack of required documents such as ownership/possession of land in the name of the farmer himself. However, many farmers alleged political biases in the selection of affected farmers by the panchayats. Many farmer households did not receive any compensation in spite of heavy crop loss by flood and rainfall. On the other hand, as it is claimed by many marginal and small farmers, the farmer households who have political power got the compensation even if no crop loss occurred to them. According to our survey, only 6.67 percent of the landless lessee cultivators and 10 percent of the landless sharecropper households received compensation for crop loss. Thus, from these data and allegations made by the truly affected farmers, it is quite clear that in many cases, economically 'backward' farmer categories having no political influence were deprived by the administration as well as by the powerful section of the farmer society in the village.

Training for Farmers

There are provisions for farmers' training under different schemes. The major objectives of these trainings are skill development among farmers, to disseminate modern agricultural technology, to promote critical inputs and agronomic practice. Major ongoing schemes under which different agricultural trainings are provided to the farmers in our study area include Seed Village Scheme, Bringing

Green Revolution to Eastern India (BGREI), National Mission on Agricultural Extension and Technology (NAMET) or Agriculture Technology Management Agency (ATMA) scheme, National Food Security Mission (NFSM). There is also a State Plan Scheme for capacity building and skill development among the farmers. Huge money is spent on the training of officials in the agriculture department as well as farmers. For example, only under the State Plan Scheme, Rs. 59.55 lakh has been utilized from 2011-12 to 2016-17 (Economic Review, 2017-18).

All the government reports measure the success of these schemes in terms of fund allotment and fund utilization. However, the physical outcome of these training programmes is not satisfactory. About 20 percent of the 360 farmer households received agriculture related training (Table-5.3). Farmers also blamed that panchayat or ADA officials do not inform or invite those farmers who are interested to get it and/or do not receive any training so far. Farmers who received such training informed us that the contents of these trainings are good enough but many of these suggestions are hard to apply due to various causes. For example, Khokan Ash, a progressive farmer of Kurkuba village, informed that many years ago farmers got the training about integrated pest management and tried to apply it for *aman* paddy cultivation. However, the total cultivated area was attacked by pest and they failed to reach the government officials in spite of repeated requests. Therefore, they did not try to apply this method of pest control again.

Subsidy for Farm Implements

For the promotion of farm mechanization, there are also many schemes. The schemes through which subsidized farm implements are provided to the farmers are One-Time Assistance for Small Farm Implements (Ota-Sfi), Financial Support Scheme for Farm Mechanization (FSSM), and Credit Linked Subsidy for Establishment of Custom Hiring Centre of Farm Machinery (CHC). Under FSSM, tractor, power tiller, power reaper, rice transplanter, pump set, rotavator etc., are provided to the farmers at the subsidy amount varying from Rs. 10,000 to Rs. 1,25,000 depending on the farm implement. Under the Ota-Sfi scheme small farm implements are provided to the small and marginal farmers at 50 percent subsidy with an upper limit of Rs. 10,000.

Under the CHC scheme, a 40 percent subsidy subject to a maximum 80 lakh per beneficiary is provided for setting up Farm Machinery Bank or Custom Hiring Centre. In 2015-16, a total of 37,624 subsidized farm implements such as tractor, pumpset, power tiller etc. had been distributed in West Bengal. Under these three schemes, a fund of about Rs. 60.31 crore has been released in 2017-18. Therefore, it is evident there is ample scope for enjoying these facilities to the farmers of our state. However, the issue is who is enjoying these facilities. During the survey, only 3.3 percent farmer households admitted that they had received a subsidy for buying farm implements (Table-5.3). However, ADA offices denied to accept this and claimed that farmer households suppressed the truth, besides many farmer households did not buy farm implements though they received a subsidy for buying this. Due to this type of cheating, at present, farmers have to buy farm implements first, then go through the process of geographical tagging and after that subsidy amount is credited to the bank account/KCC account.

If it is assumed that field data on the distribution of subsidized farm implements are unacceptable, still it provides a trend about who are getting benefitted from this scheme. Under this scheme, big farmers get the most advantage while landless lessees and landless sharecroppers are least benefitted. Possession of KCC was mandatory for enjoying subsidy under the FSSM scheme before, though it is not mandatory from 2018. Therefore, farmer households not having KCC were excluded from this scheme. Many under-privileged farmers also alleged that those who have intimacy with the ruling party members in the village get all these facilities, and they remain unaware of all these facilities or their applications are rejected. Probably, realizing this malpractice or anguish of the deprived farmers, online operating system has been started by the government. Apparently, this is a good step, but in reality resource-poor and less educated farmers rarely use the internet and receive updated information about all these programmes. Besides, they are unable to invest a large amount of money to get the subsidy under schemes like FSSM or CHC. Therefore, affluent and influential farmer households of the village mostly enjoy these government facilities.

Table 5.3
Farmers Enjoying Facilities under Different Government Schemes

Farmer household categories	Percent of farmer household received					
	Compensation for crop loss	Free Seed	Soil Testing	Subsidy for agricultural implements	Training on agriculture	None of the above
Landless lessee	6.67	13.33	0	0	20.00	73.33
Landless sharecropper	10.00	30.00	15.00	0	25.00	60.00
Marginal	32.98	32.98	27.66	2.13	21.28	44.68
Small	22.37	67.11	23.68	5.26	21.05	22.37
Medium	23.81	40.48	33.33	4.76	9.52	30.95
Big	31.58	42.11	36.84	10.53	21.05	26.32
All	27.22	40.56	26.11	3.33	20.00	39.44

Source: Field Survey, 2016-2018.

Soil Health Card Scheme

Soil Health Card (SHC) scheme was introduced by the Government of India in 2015. "SHC is a printed report containing soil status with respect to 12 parameters that a farmer will be provided for each of his holdings. A SHC will also "indicate fertilizer recommendations and soil amendment required for the farm" (Vikaspedia.in).There is provision for providing SHC to the farmers once in a cycle of three years. A sum of Rs. 190 per soil sample is provided to State Governments as cost (Vikaspedia.in). Soil samples were collected in the study area under this scheme. About 26.11 percent of the surveyed farmer households reported soil samples had been collected from their land. However, soil health cards were not distributed in time among all of those farmers.

MGNREGS and its Impact on Agricultural Infrastructure Development

MGNREGS is a scheme for rural development implemented in the year of 2005. It guarantees 100 days of employment opportunity per year to every unskilled labourer in his/her village. The daily wage under this scheme in West Bengal has increased from Rs. 170 in 2015-16 to Rs. 184 in 2019-20. This scheme has a significant impact on rural poverty reduction. Another principal aim of this scheme is infrastructure development. There is provision for works which directly or indirectly may help in agricultural development. Renovation of water bodies, digging of the pond, construction of earthen dam may prove to be helpful for irrigation; bio-fertilizer production is helpful for achieving organic farming; road construction and its maintenance, construction of flood control channel and its repairing etc. are also beneficial for agricultural development. There is also provision for the development of individual farmer's land. The works related to agrarian development done under this scheme in our study area include the renovation of ponds and irrigation channels, digging of the pond, construction and maintenance of roads, and leveling of individual's farmland. However, there is a question about the effectiveness of these implemented works. About 43.9 percent farmer households responded that no productive work had been done for agricultural development under this scheme. According to most of them,the quality of work is deficient, and a huge amount of money is wasted for all these works. They blamed that getting payment without doing an actual job has destroyed work ethics, and adversely affected labour use in agricultural activity. This issue has already been discussed in chapter 3.

Paddy Marketing by Farmers and Paddy Procurement by the Government

For providing remunerative prices to the farmers, the government organizes procurement operations for the crops covered under MSP. In 1997-98, the government started Decentralized Procurement Scheme (Press Information Bureau, 2015) to boost up the procurement of wheat and paddy and to make sure that the benefit of MSP reaches the farmers. However, the paddy procurement scheme started to regain importance in West Bengal recently, especially after the implementation of the National Food Security Mission by the central government and Rajya Khadya Suraksha Yojana by the government of West Bengal. In our study area, farmers are being benefitted by the paddy procurement scheme from 2014-15. It was hoped paddy procurement at MSP would help to reduce distress sale, and to eliminate the middleman.

However, from the study, it has been observed that only about 45.6 percent farmer households have sold paddy at MSP to the government's procurement agencies, and the rest of the farmer households (54.4 percent) have never sold their produce at MSP. Out of 45.6 percent farmer households, about 22.5 percent farmer households sold paddy to the procurement agencies only for one time in the last five years (Table-5.4). Very few percentages of farmer households had sold paddy for more than two times in the last five years. Besides, there is an inter-class variation in the distribution of farmer households by the selling of paddy at MSP. No one of the landless lessees, only 20 percent of the landless sharecroppers, and only 40 percent of the marginal farmers sold paddy to the procurement agencies. Big farmer households rank first in the selling of paddy at MSP to the procurement agencies followed by medium farmer households. Only about 15.8 percent of the big farmer households did sell paddy under paddy procurement scheme. Therefore, a question arises why so many farmer households did not sell paddy to the procurement agencies. Moreover, why there is so much variation among farmer households in case of selling paddy at MSP to the procurement agencies.

Table 5.4
Proportion of Farmers and Frequency of Selling Paddy at MSP

Farmer household categories	percentage of Farmer household sell paddy at MSP				
	0 time	1 time	2 times	3 times	4 times
Landless lessee	100.00	0	0	0	0
Landless sharecropper	80.00	15.00	5.00	0	0
Marginal	60.1	20.7	11.7	6.4	1.1
Small	44.7	23.7	25.00	3.9	2.6
Medium	35.7	31.00	21.4	7.1	4.8
Big	15.8	42.1	21.1	21.1	0
All	54.4	22.5	15.3	6.1	1.7

Source: Field Survey, 2016-2018.

The answer is that farmers faced many obstacles in selling paddy under this scheme, as well as they had other problems. During 2014-15, the number of centralized procurement centres (CPC) was very low. It caused high transportation costs for carrying paddy to the procurement centres. Though during the *kharif* marketing season of 2014-15, Rs. 15 per quintal of paddy has been provided to the farmer as an incentive for bringing paddy at CPC over the MSP of Rs. 1360 per quintal, the amount of incentive was not attractive enough. Later, government appointed rice millers, agricultural co-operatives and self help groups (SHG) as procurement agencies for buying paddy from farmers at MSP aiming more paddy procurement. However, farmers have reported that all these agencies levied an 8 percent cut on the amount of paddy. According to 19.7 percent farmer households, major problem in paddy procurement is that the procurement agency, especially rice millers, unlawfully cut very high rate of levy which resulted in lower returns than their expectation.

Farmers also need to collect coupons distributed by the agencies through panchayat. As farmers belonging in hegemony class easily collect these coupons, sometimes obtain more coupons illegally, subaltern farmers failed to do so and they remained deprived. Many farmers alleged that panchayat members or ruling party workers are receiving money from them for providing coupons. Besides, farmers need to show documents of land ownership to the agency before selling paddy. However, many farmer households, especially lessee cultivators and unrecorded sharecroppers, do not have ownership documents. Thus, they failed to grab this opportunity. More than 53 percent respondent reported that the collection of the coupon is very hard. About 80 percent of the landless lessees and 70 percent of the landless sharecropper opined they did not get coupons easily while only 15.8 percent of the big farmers believe so.

There was the provision that one farmer could sell only 25 bags of paddy weighted 60kg each during 2015-16. Therefore, many farmers failed to sell off their total surplus produce. As a result, many farmers have lost interest in this scheme. In the *kharif* marketing season of 2018, this procurement limit has been set at 150 bags of paddy per farmer (Ghosal, 2018), but a farmer can sell only upto 3 quintals of paddy at one go (Chakraborty & Sarkar, 2018). Due to this provision, many farmers are being compelled to sell their produce at a lower rate in the open market. About 40.6 percent farmer households opined that one of the major problems in the government procurement system is that they did not buy the entire paddy produced by a farmer (Table-5.5).

Another major complaint against the paddy procurement scheme is late payment by the SHG, co-operative society and rice millers. Though there was a provision that payment should be made to the farmers within three days against their sale, it was reported by the 25.8 percent farmer households that SHG, co-operatives and rice millers made late in payment for 4-6 months during *kharif* marketing seasons of 2014-15 and 2015-16. In addition, about 11.4 percent farmer

Table-5.5
Problems Related to Selling Paddy at MSP

Problems	Categories of Farmers (%)						
	Landless lessee	Landless Sharecropper	Marginal	Small	Medium	Big	All
Do not get coupon easily	80.0	70	55.3	68.4	16.7	15.8	53.3
Do not buy total produce	0	0	46.3	43.4	35.7	57.9	40.6
Delay in buying	33.3	60	15.4	17.1	9.5	5.3	17.8
Late for payment	0	30	21.8	18.4	52.4	52.6	25.8
cut high rate of levy	0	30	20.2	25.0	11.9	15.8	19.7
MSP is still low	0	10	9.6	11.8	19.0	21.1	11.4
No problems	0	0	25.5	9.2	23.8	42.1	20.3
No response	20.0	10	14.9	9.2	4.8	0.0	11.7

Source: Field Survey, 2016-2018.

households responded that the MSP of paddy announced by the government was still low and it should be more than Rs. 1600 per quintal of paddy to make the paddy cultivation profitable activity. Interestingly, medium and big farmers are demanding more MSP than small and marginal farmers.

MSP of paddy was Rs. 1360 per quintal in 2014-15. It has increased to Rs. 1550 per quintal for 2017-18 and Rs. 1750 per quintal in 2018-19. Besides, the government recently is deducting 5 percent from the paddy weight instead of 8 percent before. All these are good for farmers. However, the market price of paddy is the primary concern now as it remains less than MSP for most of the time, which leaves them with hardly any profit.

As it is told before, farmers faced troubles in getting the MSP of their paddy produced, farmers depend partly or fully on the local traders/*aratdars* to sell off their produce. From Table-5.6, it is observed that about 16.4 percent farmer households did not sell paddy to the local traders/*aratdars* during 2015-16. About 13.3 percent farmer households sold less than 50 percent of their produce to the local traders/ *aratdars*, whereas about 15.3 percent farmer households sold more than half of the produce to the local traders/*aratdars*. Therefore, it seemed that about 45 percent farmer households were able to sell paddy to the government's procurement agencies. However, this figure is less than that, because many farmer households especially landless lessees, landless sharecroppers, and marginal farmers use total paddy produced for household consumption. A percentage of medium and big farmer households also sold paddy to the rice millers directly. About 55 percent farmer households sold total amount of produced paddy to the local traders/*aratdars*, and more than half of the landless lessees, landless sharecroppers, small and marginal farmer households depend on them to sell their produce.

Table-5.6
Farmers Selling of Paddy to the Local Agents during 2016-18

Farmer households categories	Farmer household (%)				
	Did not sell paddy to the local trader/ private agent	Sold less than 50 percent of the produce	Sold more than 50 percent of the produce	Sold total pro- duce	Total
Landless lessee	20.00	0	20.00	60.00	100
Landless sharecropper	15.00	10.00	20.00	55.00	100
Marginal	13.3	18.1	10.1	58.5	100
Small	17.1	7.9	17.1	57.9	100
Medium	26.2	4.8	21.4	47.6	100
Big	21.1	21.1	36.8	21.1	100
All	16.4	13.3	15.3	55.00	100

Source: Field Survey, 2016-2018.

Selling of paddy to the local traders/*aratdars* does not necessarily mean distress selling. It is a common practice that many farmers sell part of the total amount of paddy after harvesting and hold rest of the produce hoping for market prices to firm up. Farmer households also sell paddy intermittently by a small amount to meet the cash requirement. Considering all the situations, they decide the selling of crops. They also use their previous years' experiences before selling their product. Many farmers use to sell off their total amount of paddy produced because they have lack of places like store room or yard to keep their produce at home. Therefore, it is complicated to find out what is distress sale and what is not. In the Table-5.7, farmers' behaviour regarding selling paddy has been presented. From this, it can be said that about 30.6 percent farmer households, who sell their total amount of paddy when market price remains high, are not forced into distress sale. A certain percent of farmer households among those who sell their produce instantly after harvesting and whenever cash is required are forced into distress sale.

Shrinking Primary Agricultural Credit Societies (PACS)

Co-operative society is one of the most important institutions that plays a crucial role in agriculture and rural development. Co-operative societies started to expand rapidly after the enactment of the Co-operative Societies Act in 1904 and this expansion of co-operative societies is termed as the co-operative movement. Different types of co-operative societies have been established after that. However, we would like to discuss only primary agricultural credit societies (PACS) which were formed with a cluster of villages under each block by the village farmers and artisans. It is an important pillar in the short term co-operative credit structure. During the 1980s, these PACS played a significant role in agricultural growth and rural development in West Bengal. The dominance of the informal sector in providing loans to the farmers had declined due to the presence of these PACS. Except for providing short term loan, they also supply various inputs like seed, fertilizers etc. to the marginal and small farmers. PACS helps in the distribution of Kisan Credit Card among farmers, organizing SHG and acts as a government agency in paddy procurement scheme.

There are about 20 PACS in Galsi-I CD Block and 29 PACS in Galsi-II CD Block. Society wise distributions of the marginal farmer, small farmer, agricultural labourer, borrowing members and KCC holders to total member have been provided in Appendix 5.1 and in Appendix 5.2. It is observed that more than half of the members are marginal and small farmers. Borrowing members under these societies varies from 19.5 percent to 100 percent with an average of 38.7 percent for Galsi-I CD Block and from 12.6 percent to 100 percent for Galsi-II CD Block. Besides, the distribution of KCC among the members reveals the very poor performance of the PACS. In Galsi-I CD Block, percentage

Table-5.7
Time of Selling Paddy by Farmer Households (%)

Farmer Households responses	Landless lessee	Landless Sharecropper	Marginal	Small	Medium	Big	All
Sell total surplus instantly after harvesting	46.7	55.00	44.7	14.5	11.9	0	32.8
Sell produce whenever cash is required	6.7	20.00	20.2	32.9	7.1	15.8	20.6
Try to sell total produce when the market price remains high	33.3	25.00	22.9	30.3	59.5	47.4	30.6
Sell part of produce when cash is needed and rest when market price tends to increase	0	0	3.2	14.5	16.7	21.1	7.8
Sell part of produce instantly after harvesting and rest when the market price remains high	0	0	5.9	6.6	4.8	15.8	5.8
Sell part of produce instantly after harvesting and rest whenever cash is required	13.3	0	3.2	1.3	0	0	2.5
Total	**100**	**100**	**100**	**100**	**100**	**100**	**100**

Source: Field Survey, 2016-2018.

of KCC holder member varies from 5.8 percent to 100 percent with an average of 23.7 percent. In Galsi-II CD Block, percentage of KCC holder member varies from zero percent to 98.4 percent with an average of 21.39 percent.

Moreover, many of the societies are not working at present. In Galsi-I CD Block, DVC Agril. Labour PACS and Manoharpur SKUS Limited are not working. In Galsi-II CD Block, at least 10 PACS are non-working at present. These are Belan SKUS Ltd., Gohagram Garamba SKUS Ltd., Joy Krishnapur SKUS Ltd., Kaitara SKUS Ltd., Konerpur SKUS Ltd., Masjidpur SKUS Ltd., Merual SKUS Ltd., Sanko SKUS Ltd., Satinandi SKUS Ltd., and Shyam Sundarpur SKUS Ltd. Few of the working societies are still performing well. For example, Masjidpur SKUS Ltd. and Kurkuba Askaran SKUS Ltd. have a good reputation in providing services to the farmers of surrounding villages. However, most of the societies are not functioning well compared to their past performance.Therefore, it can be said that the co-operative societies have failed to cater services to the people as per expectation and they are shrinking day by day in terms of number as well as in terms of expansion of services.

Role of Agricultural Development Department

There is a well integrated administrative set up under the Directorate of Agriculture from state to village level for implementing agricultural development schemes and look after the agricultural performance in the state.

Table-5.8
Administrative Wing under The Directorate of Agriculture

State Level	State Director of Agriculture	
District Level	Deputy Director of Agriculture (Administration)	Assisted by: 1. Deputy Director of Agriculture (W.B.P) 2. Assistant Director of Agriculture (Information)

		3. Assistant Director of Agriculture (Plant Protection) 4. Assistant Director of Agriculture (Jute)
Sub-Division Level	Assistant Director of Agriculture (Administrative)	Assisted by: 1. Assistant Director of Agriculture (Subject Matter) (Plant Protection) 2. Assistant Director of Agriculture (Subject Matter) (Crop Production) 3. Assistant Director of Agriculture (Subject Matter)(Training and Information)
Block Level	Assistant Director of Agriculture	Assisted by: 1. Agricultural Extension Officer 2. Assistant Agricultural Extension Officer
Village Level	Krishi Projukti Sahayak (KPS)	

At the Block level, Assistant Director of Agriculture (ADA) is the in-charge of agricultural development activities. At the block level, ADA is assisted by an Agricultural Extension Officer and an Assistant Agricultural Officer. Besides, there may have clerical assistants, fitter mechanic and group-D staff. After the implementation of the ATMA scheme, there is one Block Technology Manager (BTM) and one Assistant Technology Manager (ATM) in every Block level ADA office. Along with the ADA office, there are Block Seed Farms in some of the CD Blocks. Assistant Farm Manager heads this Block Seed Farm. There is such a seed farm in Galsi-I CD Block. Block level ADA offices are attached with various activities, such as implementation of agriculture related schemes at Block, agriculture related training, distribution of KCC, agricultural input business control, providing compensation for crop loss to the farmers and pension to the farmers, carrying out of field visits for various purposes, farm mechanization, soil testing and soil health card distribution, co-ordination with the agricultural allied sectors and related offices, etc. During our survey, it has been observed that there was a lack of sufficient staff for conducting all these activities smoothly. Even it has been opined that one ADA is insufficient, and more than one ADA are required for the proper functioning of these offices.

At the village level, *Krishi Prajukti Sahayak (KPS)* plays a vital role in agricultural development. One should know the origin and function of the post of KPS for understanding its importance. In 1979, a unique programme, named Lab to Land Programme, started aiming extension of Green Revolution technologies, and the post of KPS created in 1981 for building an active link between farmers and the administration and researchers. It was decided that blocks would be divided into circles, and their boundary resembles more or less with that of Gram Panchyat, and there would be one KPS for each circle. Each circle was divided into eight centres consisting of more or less 100 farm households in each. So, KPS had to deal with more or less 800 farmer households. KPS used to meet with ten progressive farmers from each circle once in two weeks, and they would provide information to the farmers and collect the same about agricultural problems. Every Friday, they had to go ADA office to submit reports on field visits. This system was known as 'Training and Visit System'. Agricultural problems which would not be solved at CD Block, were placed before the 'Fortnightly Training and Visit Meeting' held in sub-division. More complicated problems were discussed at 'Colloquium' in which administrative officers and researchers took part. Besides, researchers and experts would provide new information which was to be disseminated by the KPS to the farmers. Therefore, it can be said KPS was the 'messenger' in communication between farmers and administration.

However, there is a lack of such integrated communication due to the lack of KPS. Many of these posts remained vacant for a long time. There are only two KPS in Galsi-I CD Block and one KPS in Galsi-II CD Block. As a result, farmers are not getting any help from the ADA office when facing problems. Many farmers told us that they do not come to know the date on which seeds are sold from seed farm and other information. Besides, ADA offices are failing to provide service properly due to a lack of sufficient number of KPS. For example, the cost of cultivation data had not been collected from Galsi-II CD Block regularly as there was no KPS.

Conclusion

In conclusion, it can be said that though the central, as well as the state government, has implemented so many schemes for helping farmers, all categories of farmers are not being benefitted in similar ways. Most of the small and marginal farmers do not get any help from government schemes. Government supports are inadequate compared to the demand. Many farmer households are not aware about various government schemes. Some of the schemes are not lucrative enough to the farmers. Block level administration for agricultural development and agricultural co-operatives, which would be very helpful for the farmers, have failed to provide services up to mark. In this way, government support mechanisms have failed to provide adequate relief to the farmers.

The next chapter discusses the technology use in agriculture, with special reference of tractor use in paddy cultivation, and the problems associated with those technologies.

6 Technology Trap and the Question of Sustainable Agriculture

Introduction

To eradicate the problems related to agriculture, importance is increasingly being given to achieving sustainability rather than just increasing productivity through the use of improved technology. However, it was the adoption of 'new technology' which brought rapid agricultural growth in West Bengal State of India during the 1980s. Though it has brought about revolutionary changes in the agricultural system and offered many advantages, this modern agricultural system has completely replaced the traditional self-sufficient system, and has created a number of socio-economic and environmental problems.

There are many studies (Chand & Haque, 1998; Kumar & Reddy, 2017; Ray & Ghosh, 2007; Sarkar & Das, 2014) which show that introduction of HYV technology handicapped agriculture ecologically and caused damage to human health. It is quite an established fact that HYV technology is characterized by social inequality. At the same time, mechanization was also an important component of the Green Revolution. Modern agricultural implements like tractor, water pump and so on were introduced in place of bullocks and many traditional implements. This mechanization impacted agriculture in many ways. There are also studies showing the impact of mechanization in agriculture on production, farm income and labour. Verma (2001), based on various studies, shows that tractorization helped in increasing productivity, cropping intensity and economic upliftment.

However, the impact of tractors on labour use is debatable. Based on various studies conducted by different organizations and authors, Verma (2001) came to the conclusion that mechanization did not result in decreased labour employment; rather "...mechanization opened up new avenues for human employment such as managerial and supervisory jobs on the one hand and driving, servicing, maintenance and repair of the machines on the other" (Verma, 2001, p.151). The impact of tractorization on cattle rearing is also analyzed by various studies. Agricultural Economic Research Centre (AERC), Delhi (as cited in Verma, 2001) came to the conclusion that tractor use displaced only one pair of bullocks. As per National Council of Applied Economic Reseach (NCAER), 1980 (as cited in Verma, 2001) tractor use reduced work for draught animals, but resulted in milch stock increase. However, most of these studies were conducted during the 1960s to 1990s. Moreover, the relation between tractor use and sustainability of agriculture was not extensively discussed.

In this context, this chapter highlights how mechanization in agriculture, specifically use of tractor and harvester has created a threat to sustainable agriculture. This chapter discusses how tractor use is responsible for growing chemical fertilizer use and declining bio-fertilizer use. Besides, the environmental impact of harvester use has been addressed in brief, because harvester use for paddy cultivation is relatively new in our study area. The term 'technology trap' has been used to mean the problems related to tractor use and harvester use in agriculture. This chapter uses primary as well as secondary data. It has been attempted to portray the local scenario using state-level secondary data as there is no block-level localized secondary data.

The Technology Trap

Generally, technology helps our society in many ways. It saves cost, labour and time and makes life more comfortable. However, technology can also create problems. The problems associated with the use and benefits of technology are called the technology trap. Crabb and Stern (2010, p.19) called technology trap as a 'subspecies of social trap' which is "characterized by the use of technologies that provide immediate benefits but that pose unavoidable longer-term costs to the well-being of individual users, society and the planet". In this connection it should be mentioned that John Platt in 1973 used the term 'social trap' to mean that individuals working for short-term profit had a tendency to exploit resources, which resulted in long-term loss to the society. Therefore, Crabb and Stern (2010) define the term 'technology trap' following 'social trap'.

However, Fofack (2008) in his policy research working paper termed the low rate of technological advancement in the countries of sub-Saharan Africa compared to the rest of world as a 'technology trap'. Fofack opined that advanced economies benefitted from technological advancement and enjoy rapid income growth and rapid diversification of economy from agriculture to industry sector. On the other hand, lack of technology diffusion caused mainly by the poverty trap creates a technology trap in sub-Saharan Africa.

Crabb and Stern (2010, p.20-21) described five types of technology traps. These are:

i. **The Incompetence Trap:** It implies deskilling of users and making them dependent on technology.

ii. **The Self-miscontrol Trap:** When human behaviour is controlled by technology rather than by social norms, or humans became addicted to technology use, it is called self-miscontrol trap.

iii. **The Techno-centrism Trap:** It means over-trust and over-dependence on technology and distrust on humans.

iv. **The Misbehaviour Trap:** It denotes misuse of technology, which is against social norms and law.

v. **The Environmental Degradation Trap:** Environmental problems caused by use of technology are termed as 'technology trap'. Crabb and Stern also opined that technology-induced environmental problems are often sought to be solved using more technology, which further entraps the society.

Besides, technology may create economy-related problems also. Frey and Osborne (2013) opined that a number of jobs are at risk due to technological advancement, and predicted that a technology race is ahead that will replace low-skilled workers. Jagannathan (2018) feared that inclination of service and industrial sector companies towards automation imposed a great threat to job growth in India. Though impact of technology use on demand for human labour in agriculture is debatable, technology use does affect agricultural economy in other ways. Thus, the term '**economic trap**' has been used to show how technology use is responsible for increasing costs and for growing dependence of resource poor farmers on rich farmers.

Tractor Use and Declining Animal-Drawn Plough

There are two views regarding tractor use (Binswanger, 1978). One is the substitution view which believes that the switch from 'animal power' to 'tractor power' occures when cost for bullocks and labour becomes sufficiently high. The second view is the net contributor view, which argues that the power and efficiency of machines is higher than that of animal labour, they are able to reclaim land which cannot be operated by bullocks, the speed of tractors allows for timely operation, and thus helps to increase yield and cropping intensity without necessarily displacing labour.

Mordecai Ezekiel, an economist with the United States Department of Agriculture (as cited in Ellenberg, 2000, p.565) opined, farmers prefer tractor as it saves time and it reduces trouble of caring for horses consistently, "as well as for their greater economy on the job". A strong advocate of animal power use in agriculture, Wayne Dinsmore (as cited in Ellenberg, 2000) also admitted, "farmers were going to tractorize, wholly or in part,...to obtain ease, to escape chores, would cause a general shift, no matter how detrimental it might prove to farm interests as a whole".

In India, intervention of tractors took place mainly during the 1960s. Like other parts of India, intervention of tractors replaced animal labour use in our study area also. Tractors are mainly used for puddling and paddy transportation from the field in paddy cultivation. Due to tractor use, the number of animal-drawn wooden ploughs has decreased considerably. However, secondary data on the number of tractors per panchayat is not available. Moreover, we have been able to collect data regarding number of wooden ploughs per panchayat for the year 2003 and 2007 from the Livestock Census, which have been presented in Fig. 6.1. From this figure, it is evident that wooden ploughs are still in use in our study area though in very small numbers, and in most of the panchayats, the number of animal-drawn wooden ploughs has decreased. As there is a lack of reliable and sufficient secondary data which can help to understand the tractorization process, we relied on field data.

Table-6.1
Panchayat-wise Distribution of Wooden Plough (2003 & 2007)

Galsi-II Name of Gram Panchayat	Wooden Plough number 2003	2007	Galsi-I Name of Gram Panchayat	Wooden Plough number 2003	2007
Adra	1158	458	Budbud	484	275
Bhuri	569	485	Chaktentul	661	948
Galsi	858	84	Loa-Krishna-rampur	948	820
Gohogram	1041	951	Loa-Ram-gopalpur	843	832
Khano	435	550	Mankar	514	981
Kurkuba	82	200	Paraj	1767	1043
Masjidpur	639	488	Potna-Pursa	1279	1304
Sanko	241	129	Serorai	1278	1111
Satinandi	250	133	Ucchagram	1362	1199
Total	**5273**	**3478**	**Total**	**9136**	**8513**

Source: 17th and 18th Livestocks Census.

From the field data as represented in Fig. 6.1 and Fig. 6.2, it is clearly observed that a few sample farmer households had tractors before, while most of them used to plough their land with the help of animal-drawn ploughs. However, the situation has changed over time. It is observed that many among the sample farmer households have tractors at present, while a few farmer households use animal-drawn ploughs. Among the farmer households surveyed, only 12.78 percent have bullocks. For transportation of paddy from the field, most farmers use tractors. Only about 11.67 percent farmer households have carts at present. It has been reported

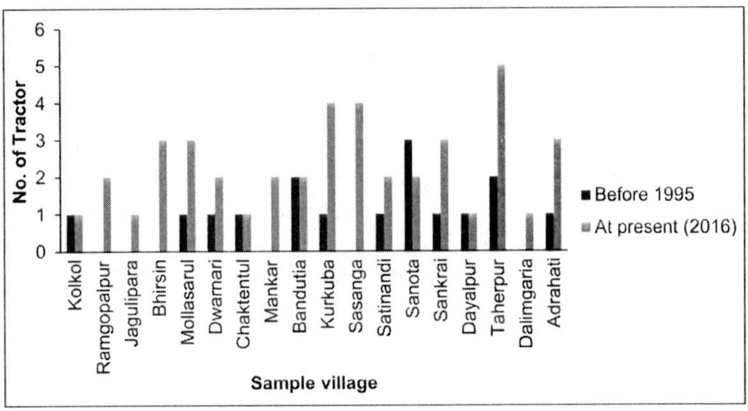

Fig. 6.1: Increase of Tractor

Source: Field Survey, 2018.

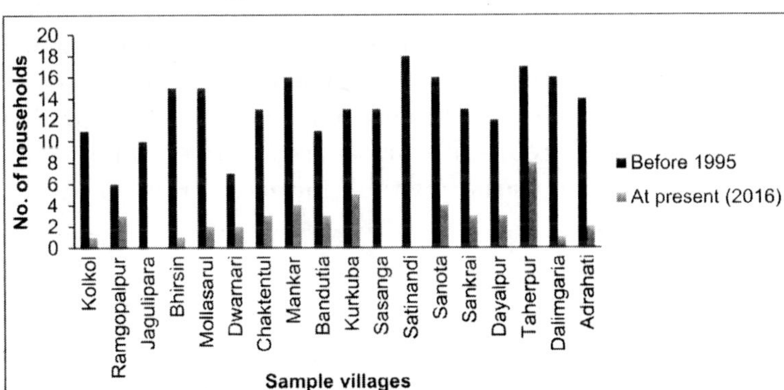

Fig. 6.2: Decrease of Animal Drawn Plough

Source: Field Survey, 2018.

that farmer households who possess animal-drawn ploughs and carts also use tractors for ploughing land and for transporting paddy from the field.

Therefore, it is imperative to know why tractor use is increasing. The most important cause for increasing tractor use is the cost advantage in tractor use. During the survey year, farmers reported that in paddy cultivation own animal-drawn ploughing costs about Rs. 2700 per acre and hired animal-drawn plough costs about Rs. 3600, whereas hired tractor-drawn plough costs only about Rs. 2000-2500. However, the cost of transporting crop either using tractor or using cart is more or less the same. Both cost on an average Rs. 500 per bigha of paddy transportation in our study area during 2018.

For completion of agricultural work within a short period of time, animal power use is not suitable. Tractors take very less time for ploughing or for transportation of harvested crop from the field. Animal-drawn ploughs take at least 12 hours to plough one acre of land one time (over 2-3 days), whereas a tractor-drawn plough takes only one hour for the same. In case of transportation of harvested paddy from field, tractors can carry paddy of 15-20 *katha* of land at one go, whereas a cart can carry the same in 4-5 rounds. As a tractor takes lesser time, farmers of our study area prefer to use tractors over carts. From Table-6.2, it is observed that about 81.94 percent farmer households totally depend on tractor for paddy transportation from field. Only 10.83 percent farmer households reported that they use only bullock carts for paddy transportation. Moreover, if a farmer uses a harvester for paddy harvesting, tractor use is inevitable for paddy transportation.

Table-6.2
Distribution of Farmer Households by Mode of Paddy Transportation from Field (%)

Farmer household categories	Tractor	Bullock cart	Tractor and Bullock cart	Cycle/ motor Van	Tractor and cycle/ motor Van
Landless lessee	0	86.67	6.67	6.67	0.00
Landless sharecropper	0	70.00	5.00	15.00	10.00
Marginal	82.98	11.17	4.79	0.53	0.53
Small	0	0	85.53	7.89	6.58
Medium	0	0	71.43	23.81	4.76
Big	0	0	0	89.47	10.53
All	81.94	10.83	6.11	0.28	0.83

Source: Field Survey, 2016-2018.

Besides, tractor-drawn ploughs are more efficient in the first ploughing for paddy cultivation. This is because at that time the soil remains relatively harder and full of grass, and ploughing using animal-drawn plough is very laborious and takes a long time. Therefore, as mentioned earlier, farmer households possessing bullocks also prefer tractor ploughing for every first ploughing of paddy cultivation.

Lack of labour for ploughing, higher wage for that labour, and rising cost for draught animal rearing is another cause for increasing tractor use. From the field data, it is evident that before 1995 about 65.56 percent and 52.39 percent surveyed farmer households had animal drawn ploughs and carts respectively. Most of these farmer households reported that as cost for maintaining draught animals rose and the amount of agricultural land per household declined, draught animal use became unprofitable compared to the use of hired tractors. As hired tractors became readily available over time, the farmers started to sell off their bullocks and use hired tractors for agricultural activities.

Many small and medium farmer households who own tractors admitted that they had bought it for timely operation as well as for their own convenience. So, Wayne Dinsmore's (as cited in Ellenberg, 2000) observation is partly applicable here. Better financial condition and availability of low-cost credit helps most of the medium and big farmer households to acquire high value machinery like tractors. "As most modern agricultural operations are tractorised, the large farmers earn much additional income by hiring their tractors out to small and marginal farmers. They also hire out the tractors during lean periods for non-agricultural use" (Mohanty, 1999, p.A58).

Therefore, it can be said that both views regarding tractor use are suitable for explaining increasing tractor use and declining animal power use in our study area. What is important to note is that growing marginalization of the farmer households in terms of land possession has also helped in tractorization in our study area.

Impact of Tractor Use on Cattle Rearing

From the interviews we learned that cattle rearing, and more specifically milch cow rearing per farmer household has decreased considerably along with the decline of draught animals. In the focused group discussion, farmers opined that earlier, farmer households were compelled to keep bullocks for various agricultural activities. They used to undertake cattle rearing, specifically milch cattle rearing, along with draught animal rearing to take advantage of scale of economies. However, with the intervention of tractors, even the farmer households who do not possess tractors started to hire tractors for agricultural activities. As bullocks became unnecessary, they were sold off. Later, to escape from chores related to cattle rearing, farmer households started to drop the number of cows they reared. In this way, tractor use indirectly affected cattle rearing. When we mentioned that during the field survey about 64.72 percent farmer households reported they have cows, farmers argued that at present farmer households keep one or two cow(s) for supply of milk and/or fuel for cooking. Our study also reveals that farmer households possessing bullocks have more than 4 cattle per households. In this connection, it should be kept in mind that there are a few villages like Dalimgariya, Adrahati, etc., where the number of cattle per farmer household is high, even though there are a few farmer households that use bullocks; these villages are dominated by *gowala* (milkmen).

It is difficult to statistically prove the impact of tractors on cattle rearing as there is a lack of required secondary data. Moreover, as farmer households failed to provide past data regarding number of cattle, we have failed to find any significant relation between decline of animal-drawn ploughs and decline of cattle per farmer household. However, from Table-6.3, it is observed that in many gram panchayats the total number of cattle has decreased between 2003 and 2007. In the state, total cattle experienced only 1.46 percent growth in number during 2003-2007, but experienced a change of 13.94 percent negative growth during 2007-2012. There may be growth in number of total cattle after 2012 due to the state's intervention. However, it should be remembered that all these figures related to cattle are not only pertaining to farmer households. All types of cattle, farmer as well as non-farmer household's cattle, cattle reared for subsistence purposes as well as for commercial purposes, are included in such a livestock census. Therefore, based on the focused group discussion, it can be inferred that tractor use has adversely impacted on milch cattle rearing.

Table-6.3
Distribution of Total Cattle in Galsi-I and Galsi-II CD Blocks, 2003 and 2007

Galsi-II Name of Gram Panchayat	Total Cattle 2003	2007	Galsi-I Name of Gram Panchayat	Total Cattle 2003	2007
Adra	4424	1405	Budbud	3710	4046
Bhuri	7798	8117	Chaktentul	8081	1469
Galsi	4612	3320	Loa-Krishna-		

			rampur	5793	6599
Gohogram	7668	9148	Loa-Ram-gopalpur	6137	12131
Khano	3276	7822	Mankar	4796	4079
Kurkuba	2677	752	Paraj	10017	6487
Masjidpur	9040	6834	Potna-Pursa	5723	5073
Sanko	3661	3865	Serorai	5623	5518
Satinandi	3928	1289	Ucchagram	13432	9769

Source: 17th and 18th Livestocks Census.

Indirect Impact of Use of Tractor on Chemical Fertilizer and Manure Use

Usually we think that chemical fertilizers have replaced manure just to raise the production, but the fact is just the reverse. It is true that subsidized low-cost chemical fertilizer and its immediate effect on crop production attract farmers. However, it is also a fact that due to intervention of tractors, animal-drawn ploughs have become obsolete. Consequently, total number of cattle per farmer household has decreased, thereby use of manure is also experiencing a declining trend over time. To compensate for lack of cow dung manure, farmer started to use more and more chemical fertilizers. As mentioned in chapter 3, though there are various bio-fertilizers, cow dung manure is the most commonly used manure. Before the introduction of Green Revolution technology, even during the 1980s-1990s, farmers used to apply sufficient amount of cow dung manure in every crop season. However, from our study it is observed that though 65.56 percent farmer households have cattle, only 51.39 percent of the total farmer households surveyed use cow dung manure at present. Moreover, farmers themselves admitted that they use insufficient cow dung manure irregularly because they do not have sufficient cattle as they had 20-25 years ago. Farmers use only 1 trolley/*bigha* of cow dung manure on average, either once every year or with a one year interval, depending on availability, whereas earlier they used to apply 1 or 2 trolleys manure per *bigha* in every season. As expected, having more cattle per farmer household, cow dung manure use is relatively higher at Bhirsin and Mankar village of Galsi-I Block and at Sanota, Sankrai, Taherpur, Dalimgaria and Adrahati of Galsi-II Block compared to the rest of the villages surveyed. Fig. 6.3 also reveals that there is a positive correlation between cattle and manure use.

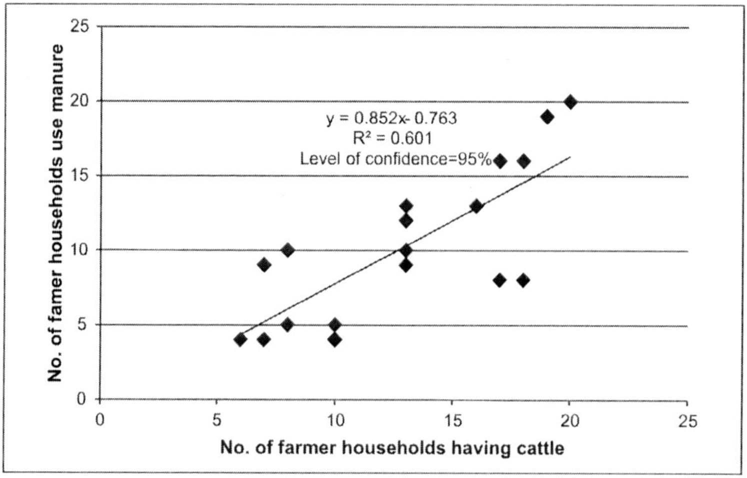

Fig. 6.3: Relation between Cattle and Manure Use

Source: Field Survey 2016-18.

Not only the primary data, but even the state-level secondary data support our claim. From Table-6.4, it is clearly evident that animal labour use in paddy cultivation of West Bengal has decreased. Simultaneously manure use has also decreased, but chemical fertilizer use has increased over time.

Table-6.4
Animal Labour use, Manure use and Chemical Fertilizer use (per hectare) in Paddy Cultivation in West Bengal

Year	Animal labour (pair hours)	Manure use (Quintal)	Chemical Fertilizer use (kg. nutrients)
1997-98	157.42	34.78	92.71
2000-01	130.6	28.02	95.81
2005-06	120.54	23.36	94.03

2009-10	79.42	24.31	150.56
2015-16	25.88	20.37	155.59

Source: Cost of Cultivation Data, Directorate of Economics and Statistics.

The relation between animal labour use and manure use (Fig. 6.4) is also positive. On the other hand, animal labour

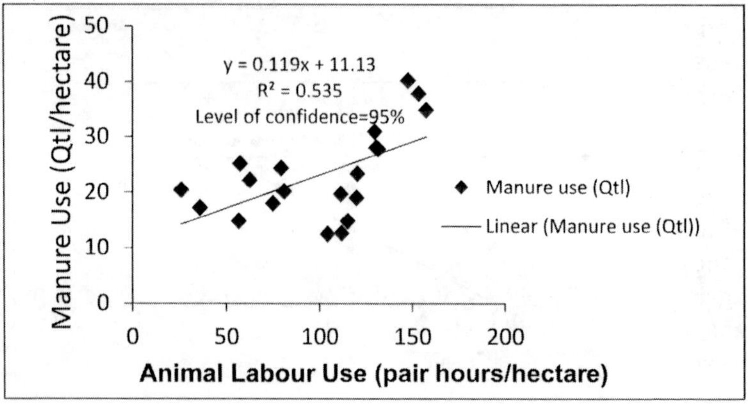

Fig. 6.4: Relation between Animal Labour Use and Manure Use

Source: Cost of Cultivation Data, Directorate of Economics and Statistics.

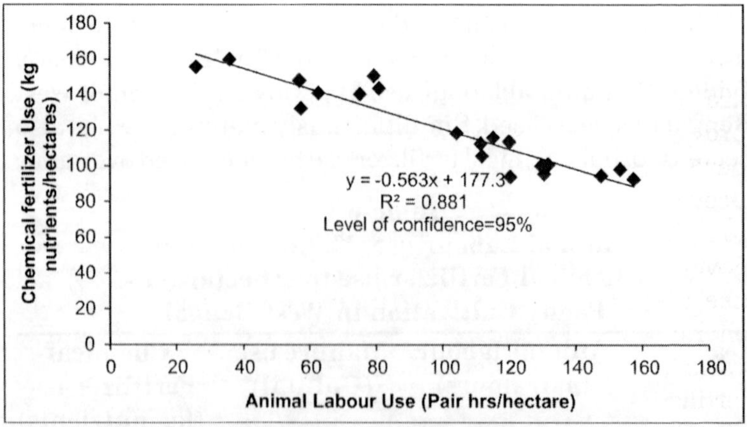

Fig. 6.5: Relation between Animal Labour Use and Chemical Fertilizer Use

Source: Cost of Cultivation Data, Directorate of Economics and Statistics.

use and chemical fertilizer use shows a negative correlation (Fig. 6.5). In both cases, calculated correlation co-efficient are significant at 0.005 level of significance.

Towards Sustainable Agriculture and the Issue of Tractorization

The present world is worried about environmental issues and thus we are looking for sustainable development. Like other sectors, the sustainability issue has emerged in agriculture sector also. The concept of sustainable agriculture is a common word in the 21st century. The concept of agro-ecology is also becoming popular, because most of the sustainable agriculture models focus on the ecological setup of a region. The Food and Agriculture Organisation (FAO) recognized agro-ecology as a key to transforming the agriculture system and achieving sustainable development goals (Pimbert, 2018). However, "sustainable agriculture seeks to address a diverse range of problems, not only ecological, but social, economic and cultural to build farming and food system that can endure in the long term" (Brown, 2018, p.3). Sustainable agriculture practice aims at lowering chemical fertilizer use, lowering reliance on external inputs.

The government is also encouraging farmers to lower chemical fertilizer use and to increase bio-fertilizer use. There are many programmes under which farmers are being provided training for vermicompost preparation. In some parts of Purba Bardhaman district interested people were provided training, vermin, and wages for vermicompost preparation under the MGNREG scheme. Staff of government offices also encouraged farmers to prepare and use vermicompost as well as bio-pesticides in place of chemicals. For paddy cultivation, application of 20-30 kg. vermicompost per *katha* can replace one-third chemical fertilizer use. However, cow dung is the main component of vermicompost. Using cow dung of one cow with 4 kg. of vermin (*eisenia fetida*), one can produce about 4 quintal of vermicompost in 3-5 months. Therefore, having no or few cattle, farmers are losing interest in vermicomposting.

Moreover, cost of vermin ranges between Rs. 1000 to Rs. 2000 per kg., which is also not cost effective for a farmer. Once more, in case of bio-pesticide and bio-vitamin, cow dung and cow urine is an essential component. Therefore, it is observed that tractorization, which caused increasing chemical fertilizer use and decline of cattle rearing among farmer households, is indirectly responsible for the threat to sustainable agriculture initiatives.

Technology Traps created by Tractorization

Therefore, it is clearly observed that tractors have many advantages over animal labour use, and are inseparable from modern agriculture practice. However, it is also observed that tractor use has resulted in environmental problems. There are studies (Chaudhary, 2018; Elaoud & Chehaibi, 2011; Krebstein et al., 2014) which show that tractor use causes

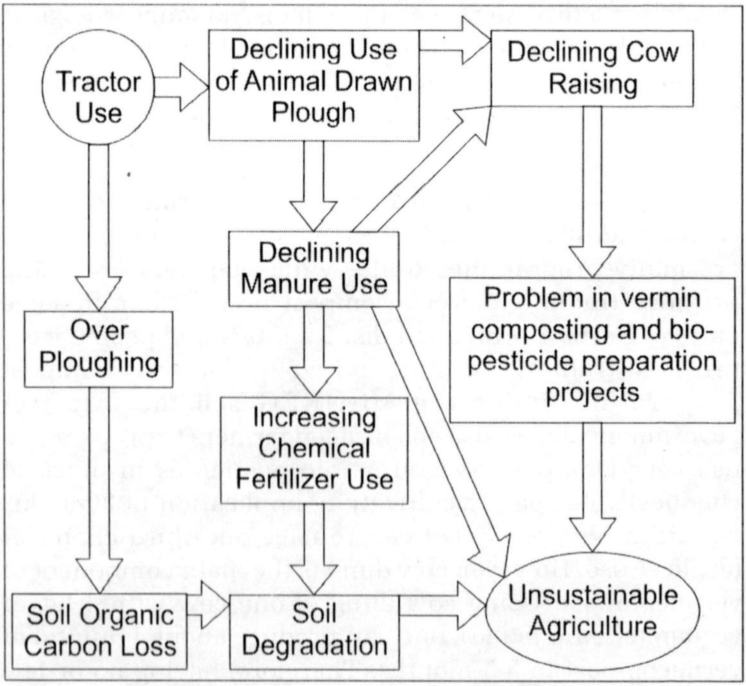

Fig. 6.6: Environmental Degradation Trap

soil compaction and over-ploughing causes loss of the soil's organic carbon. This may be termed as an environmental degradation trap, as these problems are associated with tractor use. This trap has resulted in unsustainable agriculture from the ecological point of view (Fig. 6.6).

Tractor use has also created an incompetence trap. As most farmers are now dependent on tractors for cultivation, the few farmers who still use animal-drawn ploughs face difficulties in hiring labour capable of drawing the plough. Most farmers also do not know how to draw the plough. In our study area, farmers reported that the few labourers who can use it demand high wages, and sometimes even demand part of the land of land-owner for tenancy cultivation along with the daily wage.

Moreover, relatively richer farmer households of the villages generally own a tractor. Because of the high price of tractors, a resource-poor small or marginal farmer cannot buy a tractor. From Fig. 6.7, it is evident that most owners of tractors are medium and big farmer households. A few of the marginal and small farmer households with better financial condition also possess tractors. Therefore, most of

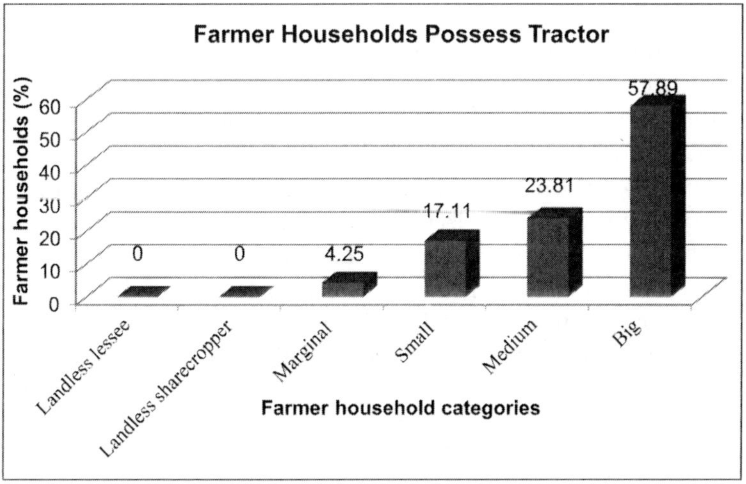

Fig. 6.7: Distribution of Farmer Households Who Own Tractor
Source: Field Survey, 2016-2018.

the resource-poor farmers have to hire tractors, and thus have to depend on the owners of the tractor, while tractor owners make a profit.

Besides, ever increasing cost for tractor use caused by increase of diesel price, increasing cost for chemicals as well as bio-fertilizers has resulted in increasing the input cost of cultivation (as it is observed that tractorization has an indirect role in increasing chemical fertilizer use and insufficient supply of manure). Due to lack of labourers who can draw the plough, non-tractor users are also compelled to pay an extra cost. During the interview with older farmers, it was opined that before tractorization, the members of the farmer's family used to do a lot of the work (plantation, harvesting and threshing etc.) along with the hired labourers. However, at present most of the farmers go to the plot only to supervise the work. Thus, it is a kind of wastage of labour in the absence of their engagement in other economic activity. All these facts can be termed as an economic trap, because these factors have resulted in input cost rise and thus are responsible for unsustainable agriculture.

Harvester Use in Paddy Cultivation and Emerging Problems

From the year 2012, farmers of our study area started to use harvesters. Within two years the use of harvesters spread rapidly. Farmers reported that rapidly increasing wages of labour and labour crisis during the time of harvesting encouraged them to use harvesters in place of human labour for paddy harvesting. Primarily, adoption of harvester was limited. Very few interested farmers would bring the harvester as well as its driver on contract either from the districts or from surrounding areas. These farmers would then hire out the harvester, look after the work, collect rent from the user farmers, and in turn get a percentage of share of total earnings from hiring out the harvester. One such medium farmer of Kurkuba village of Galsi-II Block stated that he hired a harvester from Lalgola of Murshidabad the first time and later they hired it from a *samabay samity* of Guskara. As per contract, they get Rs. 200 per hour for looking after the work and collecting due cash from the users of the harvester. The owner of the harvester bears all types of costs, except for food and lodging of the driver, helper and mechanic. Setting up of a custom hiring centre (CHC) and government subsidy has further prompted some of the rich farmers and *samabay samity* to invest in harvesters. During focused group discussion, farmers opined that harvester use has more advantages over labour use, because a harvester takes very little time and costs less, thus making harvesting very easy. It is also extremely helpful to the farmers if freak weather conditions are forecasted during the harvesting period.

The rate for harvesting paddy using a combine harvester varies depending on availability and demand of harvester, and the rate is calculated based either on time or per *bigha* of land. The rate usually ranges between Rs. 750-1200 per *bigha* and Rs. 2200 to Rs. 3500 per hour. Therefore, it is observed that a harvester takes around 20 minutes to harvest paddy from one *bigha* of land.

The adoption rate of harvesters is not the same among different categories of farmers. From Table-6.5, it is observed that most of the landless lessees, landless sharecroppers and marginal farmers did not use harvester for *aman* paddy harvesting during 2016-2018. Farmer households who have sufficient family labour and belong to financially backward sections of the society are reluctant to use harvesters. Manual labour use is more cost effective to them. Besides, the requirement of straw bundles for fuel, roof cover and fodder also compels them to use manual labour for paddy harvesting. Because of this, most farmers prefer to harvest at least part of the cultivated paddy using manual labour. Sometimes, the need for early crop harvesting requires them to use manual labour in place of harvesters. Only about 19.17 percent farmer households used harvesters to harvest the total *aman* paddy they produced.

Table-6.5
Use of Harvester for Harvesting Aman Paddy during 2016-2018

Farmer household categories	Percentage of aman paddy cropped area for which combine harvester was used in paddy harvesting					
	0%	≤ 25%	25%-50%	50%-75%	75% to <100%	100%
Landless lessee	0	0	0	86.67	6.67	6.67
Landless sharecropper	0	0	80.00	5.00	5.00	10.00
Marginal	0	82.98	3.19	1.06	3.19	9.57
Small	0	43.42	6.58	5.26	1.32	43.42
Medium	52.38	14.29	4.76	2.38	4.76	21.43
Big	0	26.32	10.53	21.05	10.53	31.58
All	68.06	5.56	3.33	0.56	3.33	19.17

Source: Field Survey, 2016-2018.

Interestingly, it is also opined by some farmers that if any one calculated the returns from straw as a by-product, then it would cost the same for both options of harvesting. According to them cutting, binding, transportation and threshing of paddy per bigha of land costs about Rs. 2500 [10 man days (4 for cutting, 2 for binding, 4 for threshing and others)* Rs. 200+Rs. 500-600 for transportation of paddy from field to farm yard]. On the other hand, renting the harvester costs Rs. 1000-1500, but harvester users do not get any returns from straw bundles for which farmers would get Rs. 600-750 per bigha. Besides, they need at least two man days to dry up the paddy, as paddy harvested by the machine retains heavy moisture which costs the farmers about Rs. 400. However, most of the farmers said harvesters reduce the cost of harvesting upto Rs. 1000-1500 per *bigha*, which is almost half of the cost incurred for harvesting by labour. Besides, harvester use will also promote a capitalistic mode of operation which may badly affect marginal farmers and agricultural labourers in the future, in the absence of a government support mechanism. Therefore, economic sustainability of harvester use is not beyond doubt.

In addition to this, heavy wastage of paddy due to harvester use is also reported by farmers of many villages. Harvester use is difficult for extremely small sized holdings. It is also not suitable for harvesting paddy from a wet land. Moreover, pollution and soil degradation due to crop residue burning is also a result of harvester use. Crop residue burning is a vital and burning issue at present, as such crop residue burning in Punjab and Haryana is an important contributing factor for air pollution in Delhi (Wikipedia, n.d.). Though there are many provisions for using crop residue from harvester use, there is lack of scope to implement those. For example, it is often recommended by scientists that crop residue could be converted into compost, as a result chemical fertilizer use would be less and soil would become more fertile. However, in a double or even triple cropped region like our study area, farmers do not get sufficient time to compost crop residue in land. Therefore, they prefer to burn it, as its alternative uses are not cost effective to them. Farmers' ignorance regarding crop residue burning is also a cause for the prevalence of this practice (Chaudhary, 2018).[16] In spite of government advertisements and news on bad effects of crop residue burning, harvester users of our study area follow this practice.

System of Rice Intensification (SRI) method and Sustainability Issues

To reduce the input cost of paddy cultivation and to ensure efficient use of resources, the System of Rice

16. While not negating the problems listed by farmers for paddy straw burning, wheat straw burning at a signicant scale clearly indicates that problems of on-farm usage of paddy are compounded by the fact that farmers do not realize the consequences of crop residue burning for the soil and, hence, for their own farm economics. Paid-out costs are easily noticed, but long-term or indirect costs are often not visible and hence, neglected (Chaudhary, 2018, p.60-61).

Intensification (SRI) method is practiced in some parts of India. This method was developed in Madagascar by Fr Henri de Laulanié.

Haldar, Honnaiah and Govindaraj (2012) in their studies conducted in Bardhaman district showed that SRI method is more efficient than the conventional system of paddy cultivation. The agriculture department and extension staffs of Galsi-I and Galsi-II CD Block also informed that they are providing training to the farmers.

However, among the sample farmer households no one has adopted this method of paddy cultivation, though many of them have heard about this method. This method is not suitable for many reasons. First of all, this method is unsuitable in low-lying areas where abundant canal or river water is available. With the increase in the distance from canals, the chances of adopting SRI method increases (Haldar et al., 2012). As many parts of our study area are located very near to the canal area, therefore the tendency for adopting SRI is very low. Besides, this method needs larger number of skilled labourers and more manure than the conventional method. Issues related to lack of sufficient labour and sufficient manure use have already been discussed. Last but not the least is the lack of a unified decision regarding use of modern technology or methods, and lack of co-operation from the neighbouring farmers which also prevents farmers from applying this method at a larger scale. It is difficult to follow SRI method when the owners of the surrounding holdings practice traditional method of paddy cultivation.

Even so, as this method requires less water and less chemical fertilizer, it is ecologically more sustainable. Various studies related to cost efficiency of SRI method (Haldar et al., 2012; Durga & Kumar, 2013; Mithra & Bhaskaran, 2018) show that this method fetches more returns to the farmers, making it economically sustainable as well. Moreover, as the canal irrigation system of our study area has gradually been losing its efficiency, this method will be suitable for our study area, specifically for *boro* paddy cultivation. However, skill development among farmers through proper training and supervision of the extension staff is crucial for the success of this method.

Zero Tillage Machine

Very recently, another machine has emerged in Purba Bardhaman district that could contribute to making paddy cultivation ecologically as well economically sustainable. The name of this machine is the Zero Tillage Machine (Zero Till Seed cum Fertilizer Drill). This machine is mainly operational in Haryana and Punjab, and is new to the farmers of West Bengal. This machine "helps farmers to seed a crop directly into the cultivated field just after the harvest of the previous crop with the least disturbance of the soil" (Yadav et al., 2002). This machine allows for soil and water conservation through the least disturbance of soil, less fertilizer and water requirement. We were present when an agricultural farm under the ADA of Galsi-I Block was using this machine for a demonstration of paddy seed plantation on their own agricultural land during *boro* season in 2017. However, proper training for the farmers is essential to use this machine properly. Besides, this machine is also costly and needs a tractor with more power (Gill, 2018).

Conclusion

Our analysis thus argues that machines used in agriculture have some negative consequences on the environment as well as on society. These consequences have resulted in agriculture tending towards unsustainability. However, it is also observed that various technologies focusing on environmental and economic sustainability are available. Agricultural institutions are now also emphasizing the importance of sustainability. Therefore, government initiatives for educating farmers about modern technologies and sustainable agriculture are needed. The incentives for adopting environment-friendly technologies and machines are required. Farmers' awareness level should be enhanced on the harmful impacts of chemical fertilizer on environment, as well as on human health. Steps should be taken so that producers of organic or natural agricultural products get proper markets for their produce. There is also a need for steps to increase per-household cattle rearing. Innovation of low-cost environment friendly machines and tools are also required. This would help the marginal and small farmer households to achieve economic sustainability as well as to achieve agricultural sustainability. Moreover, there is an urgent need for a movement mobilizing the farming society for a change from the ongoing Green Revolution technology based agriculture practice to sustainable agriculture practice.

7 Changing Lifestyle and Increasing Distress

Introduction

In the previous chapters, we have identified several problems related to agricultural practices in Galsi-I and Galsi-II CD Blocks. Most of these problems are similar to the agricultural problems of the other paddy growing areas of West Bengal, whereas some problems are identical all over India. Few problems are confined only in the study area. All of these problems related to agriculture have profound implications on the farmer society and their livelihoods.

However, agriculture cannot be the sole factor of socio-economic development or socio-economic changes in a farming society. Societal change is a normal process guided by various factors. "There are no theories generating models of sequential change and development at the village level which are theoretically persuasive" (Adelman & Dalton, 1971, p.563). In the context of changes in village society and agrarian economy after the 1990s, Jodhka (2016, p.7) opined that "the contemporary rural also represents empirical puzzles and conceptual challenges". In his words (Jodhka, 2016, p.7):

> Growing instances of farmer suicides are not simply a reflection of declining incomes and rising indebtedness among the cultivators. They are also a reflection of a complex social change and a rapid disintegration of local communities and social networks. Caste, land and economic disparities continue to matter but they can no longer be understood in the framework of *jajmani* relations, as the text book view of village social and

economic life would suggest. The Indian village today is also witness to an aspirational revolution and the desire to move out of agriculture is likely to grow as the pace of its integration into the regional, the national, and the global grows.

In this backdrop, this chapter tries to show the socio-economic and cultural changes that have taken place in the farmer society in the study area over the last twenty years or more. It also attempts to show how social and cultural change has increased the expenditure burden of farmers' households which cannot be met by the limited returns from agriculture. Major problems of agriculture, as well as farmers' society as perceived by the farmer households, have also been mentioned in brief.

This chapter is based on primary data collected from the field survey. Besides household survey data, information gathered from key informant survey and focus group discussions have also been used. A 3-point Likert-type scale has been used to measure the perceptions of farmer households on change in the standard of living. Percentage analysis has been used for the interpretation of field data.

Farmer Society: An Overview

In any socio-economic study, caste, education, economic status, housing and household amenities are crucial aspects of understanding a society. Therefore, before discussing the changes in farmer society, it is pertinent to briefly describe farmer society.

Caste

Caste is an important factor for farmers' development. It is observed that "farmers from the socially-marginalized castes hardly benefited from accessing the extension services" (Krishna et al., 2019, p.1). Among the agricultural households in India, 13.44 percent are ST, 16.26 percent are SC, 45.43 percent are OBC, and 24.87 percent belong to other castes. The figures of West Bengal differ from the national scenario.

Among agricultural households in West Bengal, 7.47 percent are ST, 26.47 percent are SC, 15.71 percent are OBC, and 50.34 percent belong to other castes. Therefore, it is found that other castes comprise the majority of agricultural households in West Bengal, whereas most of the agricultural households in India belong to OBC (NSSO 70th round, 2013).

In Galsi-I and Galsi-II CD Blocks, out of 360 farmer households surveyed, about 46.11 percent belong to the General category followed by Other Backward Class (OBC) (about 32.22 percent) and Scheduled Caste (SC) (about 20.83 percent). Only 0.83 percent farmer households are Scheduled Tribes (ST). A low concentration of ST in the surveyed villages is the reason for the low percentage of ST among farmer households. A definite pattern is observed from the distribution of farmer households by caste (Table-7.1). Most of the landless lessees (80 percent) and landless sharecroppers (55 percent) are SC, whereas most of the medium and big farmer households belong to the General caste. OBCs represent a considerable share in all farmer categories, except in landless lessees.

Table-7.1
Distribution of Farmer Households by Caste (%)

Farmer household categories	Gen	SC	ST	OBC
Landless lessee	6.67	80.00	0.00	13.33
Landless sharecropper	15.00	55.00	0.00	30.00
Marginal	45.21	21.28	1.60	31.91
Small	52.63	6.58	0.00	40.79
Medium	61.90	11.90	0.00	26.19
Big	57.89	10.53	0.00	31.58
All	46.11	20.83	0.83	32.22

Source: Field Survey, 2016-2018.

Level of Poverty

The percentage of households belonging to the above poverty line (APL) or below poverty line (BPL) is an important indicator in determining economic development of a society. Based on the list of APL/BPL, different types of facilities are provided to the BPL families through the public distribution system (PDS). As per the NSSO 70th round survey (2013), in West Bengal, about 31.7 percent of the agricultural households possess BPL ration cards, and 3.1 percent possess *antodaya* cards. In India, a larger percentage of agricultural households possess BPL ration cards (36.4 percent) and *antodaya* cards (4.9 percent).

In the study area, about 23.33 percent farmer households possess BPL ration cards (Fig. 7.1). Though NSSO data are not comparable with the field data, it can be said that the average financial condition of the farmer households in the study area is better than the farmer households of the state and of the nation. However, there is a considerable variation in the size category-wise distribution of farmer households by BPL cardholders. Most of the landless lessee households (80 percent) and landless sharecropper households (65 percent) belong to the BPL category. Among the marginal farmer households, about 25 percent are BPL. Thus, it is observed that the prevalence of poverty has declined with increasing farm size (Kumar, Kumar & Sharma, 2011).

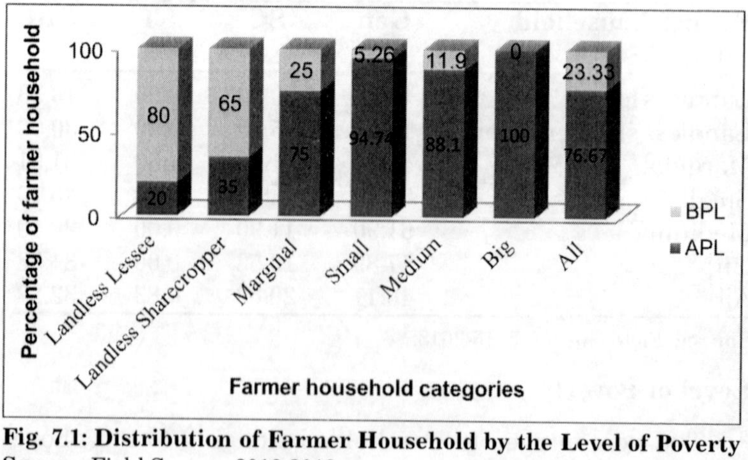

Fig. 7.1: Distribution of Farmer Household by the Level of Poverty
Source: Field Survey, 2016-2018.

During the field survey, many respondents reported malpractice in preparing the APL/BPL list. They stated that many well-off families, having proximity to the ruling party and nexus with the ration dealers and civic police, who have enlisted their name in the BPL list, while the families who really deserve to be in the BPL list have been enlisted in the APL list. This allegation may be true to some extent. Interestingly, only 5.26 percent of the small farmer households possess BPL ration cards, whereas about 11.9 percent of the medium farmers possess BPL ration cards. This anomaly somewhat establishes the allegation of malpractice in distributing BPL ration cards. However, there is no denial on the fact that poor people are over-represented in landless lessee and sharecropper categories.

Level of Formal Education

> Education is important to the improvement of agricultural productivity such that formal education opens the mind of the farmer to knowledge, non-formal education gives the farmer hands-on training and better methods of farming and informal education keeps the farmer abreast with changing innovations and ideas and allows farmer to share experience gained.
>
> Eric, Prince & Elfrada (2014, p.1951)

From Fig. 7.2, it is observed that the level of formal education of the farmers is not very high. Only 21.39 percent farmers are educated above the secondary level. About 41.67 percent farmers are educated up to the secondary level. Illiteracy is higher among the landless lessees and landless sharecroppers. On the contrary, a higher level of education is observed among the medium and big farmers. In fact, highly educated persons in West Bengal do not like to be engaged in cultivation. Instead, they prefer to be involved in non-farm activities because agriculture is considered as a hard and less prestigious economic activity. Therefore, the level of education among farmers is not high.

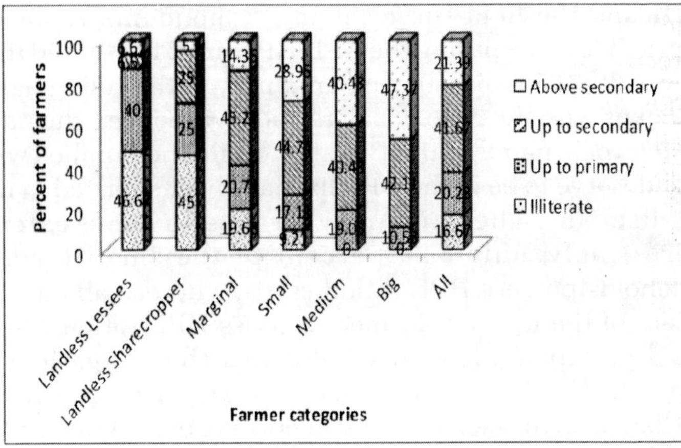

Fig. 7.2: Level of Education of the Farmers

Source: Field Survey, 2016-2018.

However, most of the farmer households have informal and non-formal knowledge of agriculture. The TV is one of the most important non-formal sources of agricultural knowledge to the farmers, as 55.6 percent farmer households informed that they watch agriculture-related programmes such as *annadata*[17], or news of weather forecasting, and apply that knowledge in agriculture. Radio is the second most important source of knowledge to the farmers. However, the importance of radio as a source of agricultural knowledge to the farmer has decreased, whereas the importance of the internet is increasing. Small and medium farmers are far ahead than the other categories of farmer in terms of gathering knowledge and information related to agriculture from TV, radio, or other sources (Table-7.2).

Table-7.2

Distribution of Farmer Households by the Sources of Agricultural Information to the Farmers (%)

Farmer household categories	Radio	T.V	Radio+ Internet	Radio+ T.V	T.V+ Internet	None of these
Landless lessee	0	26.7	0	0	0	73.3

17. A TV agricultural programme beamed by ETV.

Landless sharecropper	0	50.0	0	10.0	0	40.0
Marginal	4.8	54.3	0.5	2.7	0.5	37.2
Small	2.6	67.1	2.6	5.3	2.6	19.7
Medium	0.0	64.3	11.9	2.4	2.4	19.0
Big	10.5	42.1	0	10.5	0	36.8
All	3.6	55.6	2.2	3.9	1.7	33.1

Source: Field Survey, 2016-2018.

Housing and Household Amenities

A proper house, separate kitchen, toilet, and drinking water are the basic needs of any family for leading a decent life. Besides, in modern times every family desires to have a TV, a bike, a mobile phone, etc. All these reflect a household's standard of living. As per the NSSO 70th round survey (2013), among agricultural households in West Bengal, 12.9 percent have *kacha* houses, 44 percent have *pucca* houses, and 43.1 percent have semi-*pucca* houses. In India, among agricultural households, 9.9 percent and 61.7 percent have *kacha* and *pucca* houses respectively.

In the study area, about 47.78 percent have *kacha* houses, 38.89 percent have *pucca* houses, and the rest of the farmer households have mixed type houses (Table-7.3). Most of the landless lessee, landless sharecropper, and marginal farmer households have *kacha* houses whereas more than 50 percent among the small, medium and big farmer households have *pucca* houses. The type of house also indicates the financial condition of households. Therefore, it can be said that due to poor financial conditions most of the landless lessees, sharecroppers and marginal farmer households are compelled to reside in *kacha* houses. However, many farmer households reside in *kacha* or mixed houses in spite of having better financial conditions as their *kacha*/mixed houses are in good condition. Besides, the high percentage of *kacha* houses in the study area indicates lack of government support (Pradhan Mantri Gramin Awaas Yojana) for housing in the area.

Table-7.3

Distribution of Farmer Households by House and Different Household Amenities

	Farmer household categories (%)						
	Landless lessee	**Landless Sharecropper**	**Marginal**	**Small**	**Medium**	**Big**	**All**
			House type				
Kacha	80	60	58.51	34.21	23.81	10.53	47.78
Pucca	13.33	15	29.79	53.95	57.14	73.68	38.89
Mixed	6.67	25	11.7	11.84	19.05	15.79	13.33
			Sufficient room				
Have	13.33	40	61.17	75	85.71	94.74	65.56
Have not	86.67	60	38.83	25	14.29	5.26	34.44
			Separate kitchen room				
Have	40	50	54.26	80.26	85.71	94.74	64.72
Have not	60	50	45.74	19.74	14.29	5.26	35.28
			Toilet				
Have	66.67	85	87.23	85.53	95.24	100	87.5
Have not	33.33	15	12.77	14.47	4.76	0	12.5
			Sources of drinking water				
Own tubewell	26.7	20	56.4	72.4	83.3	57.9	59.7
Public/neighbour's tubewell	73.3	80	40.4	19.7	2.4	5.3	33.3
Own submersible	0	0	3.2	7.9	14.3	36.8	6.9

Source: Field Survey, 2016-2018.

About 65.56 percent farmer households have sufficient number of rooms in their house, and as per expectation, most of the landless lessees and sharecroppers do not have adequate rooms in their houses. On average, about 64.72 percent farmer households have a separate kitchen room in their house. Nearly 50 percent among the landless lessee, landless sharecroppers and marginal farmer households do not have a separate kitchen in their houses, whereas more than 80 percent of the small, medium, and big farmer households have a separate kitchen room in their house.

About 87.5 percent farmer households in the study area have a sanitary toilet. This figure is quite impressive as, during 2014-15, only 65 percent of the rural households in West Bengal had sanitary toilets (NSSO, 2017). Government initiatives for setting up toilets in each and every household at low cost is a major factor behind such an impressive growth in households having toilets in our study area. However, about 33.33 percent of the landless lessee households do not have toilets, and interestingly, two medium farmer households do not have toilets. Lack of consciousness may be the cause of the medium farmer households not having a toilet in the house.

Changes in Consumption Behaviour of Farmers' Households

Considerable behavioural changes in consumption pattern have been taken place among rural households including farmer households. These behavioural changes not only bring about social changes but have also propelled the economic structure of farmer society. We would like to discuss the behavioural changes related to possession of television and bike, use of fuel for cooking, use of rice for cooking, use of puffed rice, and consumption of edible oil. Generally, possession of assets like TV, bike, mobile phone and gas connection are correlated with income growth. However, "growth of these assets and incomes are hardly correlated at all" (Harttgen et al., 2011, p.16). Therefore, possession of bike and TV is considered as a behavioural change.

Possession of Television and Motorbike

The possession of television (TV) before 1995 may be equated with the financial condition of rural households. However, it is a part of the normal living conditions at present. Before 1995, as per field survey, only 18.61 percent of the farmer households had TV, whereas only 19.72 percent farmer households do not possess a TV at present. Before 1995, no one among the landless lessees and landless sharecroppers had TVs, but at present 73.33 percent of the landless lessees and 65 percent of the landless sharecropper households possess TV. The possession of TV among other classes has also increased over time. However, more than half of the big farmer households had a TV even before 1995 (Fig. 7.3).

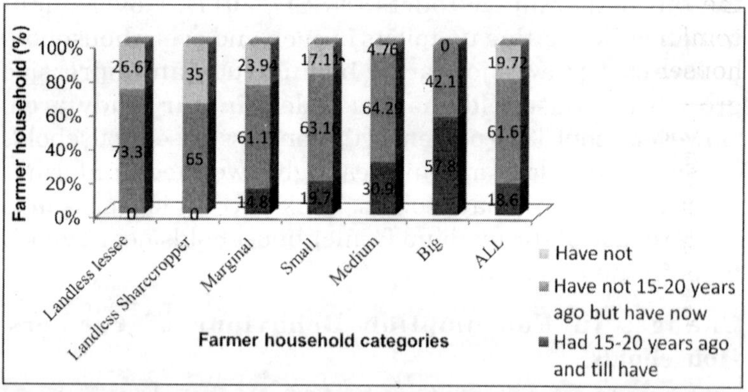

Fig. 7.3: Distribution of Farmer Households by Possession of TV
Source: Field Survey, 2016-2018.

Farmers of the study area opined that TV was considered as a highly luxurious good before, and only affluent households in the village used to buy a TV. However, due to electrification among the rural villages, trade liberalization and spread of consumerism, TV is now a popular and indispensable household commodity. A colour TV with cable connection is considered a necessary means of entertainment for a family nowadays. Farmer households consider it an ordinary expense to spend money every month for a cable connection or electricity consumption of the TV. Thus, it can be said that possession of TV even by the impoverished farmer households is one kind of asset drift.[18]

The same thing is happening in the case of possession of a motorbike (Fig. 7.4). The possession of bikes among rural households is also increasing rapidly. Though about 62 percent farmer households do not possess a bike, it is one kind of status symbol in the rural area at present. Indeed, a percentage of farmer households need a bike for business or commuting, but these are not the sole causes of possessing a bike. The young generation in the villages prefers to ride a bike rather than a bicycle. It has also been observed that farmer households with meager monthly income have the bike and use it as complementary to public transport, specifically to the bus service, to make transportation easier and more comfortable. However, we could not find any landless lessee households possessing a bike among the sample households.

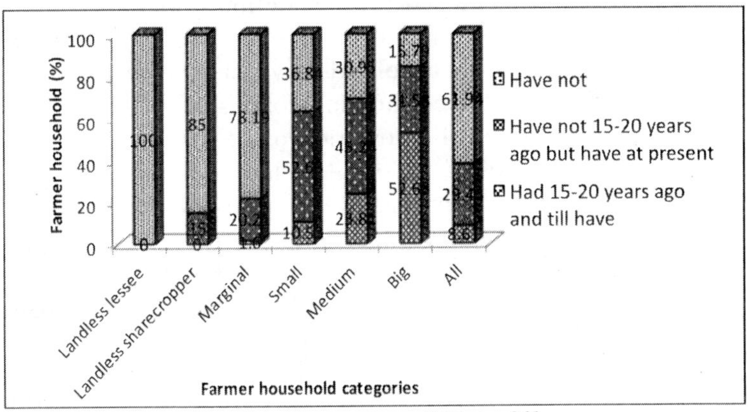

Fig. 7.4: Possession of Motorbike

Source: Field Survey, 2016-2018.

Fuel for Cooking

Another significant change in rural India is the use of gas for cooking. Before, most of the farmer households were accustomed to using cow dung, wood and crop residue as a

18. Accumulation of assets at the household level even in the absence of income growth (Harttgen, Klasen & Vollmer, 2011).

source of fuel for cooking. However, due to government schemes and availability of low cost subsidized LPG, the use of cooking gas is increasing. Being smoke-free and easy to use, most of the females prefer to use gas in place of cow dung and wood. NSSO data reveals that all other fuel types registered a declining trend during the period 2001 to 2011, whereas LPG use has increased in both urban and rural areas (Patra, 2012).

Kerosene was a popular source of fuel before, but now due to increasing cost and unavailability of adequate kerosene to all ration card holders from PDS (public distribution system), its use has decreased. Among the farmer households nearly 59 percent farmer households have LPG connections. Approximately 18.61 percent farmer households use only gas for cooking. About 40 percent farmer households use gas as well as conventional sources of fuel for cooking (Table-7.4). The study reveals that most of the farmer households use other sources of fuel along with LPG to save the relatively higher cost of gas. Though rural households have positive perceptions regarding LPG fuel, the connection cost as well as fuel cost are major obstacles to the adoption of LPG fuel (Gould & Urpelainen, 2018). All of the landless lessee households, 75 percent of the landless sharecroppers and about 46.81 percent of the marginal farmer households do not use LPG for cooking. All these farmer households are not financially well-off. The strategy followed by farmer households to save the cost of gas also reveals their financial limitations.

Table-7.4
Sources of Fuel Use by Farmer Household

Farmer household categories	Percentage of farmer household use		
	Only Gas	Gas+Others	Others
Landless lessee	0	0	100.00
Landless sharecropper	0	25.00	75.00
Marginal	15.96	37.23	46.81
Small	25.00	47.37	27.63
Medium	28.57	52.38	19.05
Big	31.58	57.89	10.53

All	18.61	40.00	41.39

Source: Field Survey, 2016-2018.

Cooking Rice and Puffed Rice

Farmer households in rural areas used to be commonly engaged in hulled rice processing to get the required boiled rice and *muri* rice.[19] Apart from household consumption, farmers used to give boiled rice as a part of wages to the labourers. However, this age-old practice of procuring boiled rice and *muri* rice at home is gradually becoming obsolete. "With the emergence of efficient modern mills, the importance of traditional hullers and shellers became redundant" (Singha & Lokesh, 2012, p.31). From Table-7.5, it is observed that about 48.89 percent farmer households have been processing the total amount of required boiled rice for consumption at home for a long time. About 27.78 percent farmer households used to process rice at home before 1995, but at present buy it from the market. A percentage of farmer households procure part of the required rice at home, and the rest of the rice from various sources. Most of the big farmers have been buying boiled rice from the market for a long time. Therefore, dependence on the market for boiled rice has increased over time. Change in the source of rice for household consumption is more pronounced among the medium farmer households, followed by marginal and small farmers. Contrary to this prevailing trend, 2.78 percent farmer households did not procure rice at home earlier, but they do so at present. Many poor farmer households also buy rice from the ration shop.

In the case of puffed rice also, most of the rural people have abandoned the practice of producing *muri* rice at home. As per survey data (Table-7.6), only 14.4 percent farmer households still produce *muri* rice, as well as puffed rice at home. About 25.6 percent farmer households buy *muri* rice

19. In this chapter, puffed rice is considered as *muri*, whereas *muri* rice is the rice used for puffed rice making. Processing of boiled or para boiled rice and *muri* rice is different.

Table-7.5

Distribution of Farmer Households by the Sources of Boiled Rice

Rice for cooking	Farmer Household Categories (%)						
	Landless lessee	Landless Sharecropper	Marginal	Small	Medium	Big	All
Have been processing at home for long time	66.67	40.00	52.66	55.26	26.19	31.58	48.89
Used to process at home before but not at present	0	15.00	29.79	26.3	40.48	21.05	27.78
Have been buying from market	13.33	0	13.30	13.16	26.19	47.37	15.83
Used to buy before but process at home now	6.67	10.00	2.13	0	7.14	0	2.78
Have been processing a part at home and getting the rest from ration/rice as wages	13.33	0	0	0	0	0	0.56
Get from ration shop/ rice as wages	0	15.00	0	0	0	0	0.83
Have been processing a part at home and buying the rest from market	0	20.00	2.13	5.26	0	0	3.33
Total	**100**	**100**	**100**	**100**	**100**	**100**	**100**

Source: Field Survey, 2016-2018.

from the market and then process puffed rice at home. Most of the farmer households (41.4 percent) buy *muri* rice from the market, then wash it, add salt and dry it at home, and finally get it fried at the *muri* factory by paying Rs. 3-5 per kg. of rice. This practice is followed by most of the farmer households because it is cost-effective, as well as healthier than the directly purchased puffed rice from the market. About 18.6 percent farmer households directly purchase puffed rice either from *muri* factories or from grocery shops.

Table-7.6
The Practice of Procuring Puffed Rice

Farmer household categories	Farmer households			
	Process *muri* rice and puffed rice at home	Buy *muri* rice, and make puffed rice at home	Buy *muri* rice, and make puffed rice from factory	Buy puffed rice
Landless lessee	6.7	26.7	33.3	33.3
Landless sharecropper	20	25	35	20
Marginal	14.9	31.9	41	12.2
Small	17.1	17.1	42.1	23.7
Medium	7.1	19	40.5	33.3
Big	15.8	10.5	57.9	15.8
All	14.4	25.6	41.4	18.6

Source: Field Survey, 2016-2018.

The major causes for the declining trend of procuring boiled rice or puffed rice at home among the farmer households, as revealed from the interviews, are as follows:

i. Crisis of skilled labour for boiled rice or *muri* rice and puffed rice processing at home.
ii. Such processing required hard labour and adequate time from the members of farmer households.
iii. Members of the family are reluctant to be engaged in

this activity, as it is either too labourious or family members do not know the process.

iv. The cost of processing for these items at home is either equal to or more compared to milled rice or puffed rice.

v. They are busy in other activities, and thus family members have no time for work related to rice or puffed rice processing.

The dependence on the market for boiled rice and *muri* has also resulted in an increase in household expenditure. Farmers told us that as family members used to process rice and *muri* at home, they had to pay only a minimum cost for hulling rice from traditional mills, but at present they sell off paddy and buy rice (boiled rice and *muri* rice) at a high rate from the market. The price of rice (boiled rice and *muri* rice) is also increasing rapidly compared to the price increase of paddy, resulting in extra expenditure.

Edible Oil

Most of the farmers used to produce mustard and/or sesamum before, at least for meeting the household consumption demand for edible oil. *Ghani*[20] was used for producing edible oil in the rural areas or the local market. This practice of producing mustard/sesamum and procuring oil from it for household consumption is also losing its importance after liberalization, because mustard production in India has become less competitive due to import of low-cost oilseeds from countries like Malaysia (Bhalla, 1999).

It is reported that about 45.28 percent farmer households in the study area have been buying mustard from the market for the last 20-25 years. Interestingly, about 22.5 percent farmer households said they used to consume *ghani* oil before, but they are buying mustard oil from the market now. Furthermore, though about 33.3 percent farmer households cultivate mustard, only 26.94 percent farmer households use *ghani* oil processed from the mustard produced by them. So,

20. A traditional machine for extracting edible oil from oilseeds.

6.34 percent farmer households buy oil even though they cultivate mustard.

Causes for growing dependence on the market for edible oil, other than the import of low priced oilseeds, are as follows:

i. Due to the prevalence of paddy and potato cultivation, mustard or sesamum cultivation per household has declined. Besides, farmers have reported that due to soil fertility loss, root swelling of mustard sapling is very common, which affects productivity.
ii. Post-harvesting procedure for producing oil from mustard requires time as well as labour. Thus, families with a single adult male member are reluctant to engage in this process. Therefore, many farmer households sell off the mustard that they produce and buy edible oil from the market.
iii. The price of marketed edible oil is cheaper than the cost of oil produced from local *ghani*. The primary cause of the price difference is that local *ghani* produces 5-10 percent less oil than the modern oil expelling machine. Besides, farmers opined that marketed oil is not as pure as *ghani* oil.
iv. Due to steep competition with modern machines, many traditional *ghani* are not operational at present. Thus, far away locations and closing up of local *ghanis* have compelled farmers to become more dependent on the market.
v. Change in edible oil consumption pattern is also responsible for growing market dependency. Nowadays, people not only consume mustard oil, but also other types of edible oil such as soyabean oil, palm oil, sunflower oil, and rice bran oil.

Change in Standard of Living and Household Savings

Behavioural changes of the farmer households are not stand alone case; rather those are associated with the changes in standard of living. We did not try to measure the quality of life using various statistical methods, but made an effort to understand the change in standard of living over the period of two decades based on the perceptions of the farmer households. Irrespective of farmer household categories, most of the farmer households opined that their standard of living has improved. Only 5.3 percent farmer households opined it has worsened, while about 20 percent farmer households responded that standard of living remained more or less the same over time. A point to be noted is that the landless sharecroppers and marginal farmers reported worsening or no change in the standard of living over time (Table-7.7).

Table-7.7
Perception of Farmer on the Standard of Living

Farmer household categories	Farmer households opined on standard of living (%)			
	Improved	Worsen	Remained same	No Response
Landless lessee	93.3	0	6.7	0
Landless sharecropper	60	10	30	0
Marginal	56.4	7.4	30.3	5.9
Small	90.8	1.3	6.6	1.3
Medium	85.7	4.8	4.8	4.8
Big	94.7	0	5.3	0
All	70.8	5.3	20.0	3.9

Source: Field Survey, 2016-2018.

However, about 60 percent of the farmer households opined that the ratio of savings to income per family has decreased compared to the time of pre-1995 (Table-7.8). Only about 22.8 percent farmer households opined that the ratio of savings to income per family has increased, and among those farmer households, most are big farmers (47.4 percent) and medium farmers (35.7 percent). Aged farmers opined that 'luxury' has resulted in rapidly increasing monthly expenditure, but income per family has not increased at such a pace, which has further resulted in decreasing savings. Here, 'luxury' as perceived by the aged persons can be equated with the improvement of standard of living. However, during focus group discussions, young or middle-aged farmers and members of the farmer households opined that standard of living has changed mainly because of education and spread of urban culture in rural areas, and the major causes for decreasing household-level savings are the price inflation, uncertain and low returns from cultivation, and increasing medical and educational expenditure.

In this connection, it should be mentioned that the standard of living of most of the farmer households is not good enough, and there exists a gap "between their actual income and income levels to sufficient to ensure a decent standard of living" (Gneiting & Sonenshine, 2018).

Table-7.8
Farmer's Opinion on the Change of Household Savings

Farmer household categories	Farmer households opined ratio of savings to income has			
	Increased	Decreased	Remained the same	No response
Landless lessee	6.7	86.7	6.7	0
Landless sharecropper	15.0	75.0	10.0	0
Marginal	21.8	57.4	16.0	4.8
Small	17.1	69.7	10.5	2.6
Medium	35.7	50.0	9.5	4.8
Big	47.4	31.6	10.5	10.5
All	22.8	60.0	13.1	4.2

Source: Field Survey, 2016-2018.

Major Problems of Agriculture and of Farmers as Perceived by the Farmer

It is already discussed in length and breadth that, there are many problems in agriculture which are responsible for farmers' distress all over India. However, farmers were asked to mention the major problems of present-day agriculture as they perceive it. From the study, it is observed that low price of paddy is the most important agricultural problem for the farmers, as opined by about 81.11 percent farmer households (Table-7.9). The second most crucial problem is the increasing cost of inputs of crop cultivation. About 75 percent farmer households opined that input costs, specifically the cost of chemical fertilizer and hired labour has increased too fast during the last ten years. As a result, profit margin has declined.

Another important problem as perceived by the farmers is that farmers do not get the actual price of their products. According to farmers, they work hard and take all the risks involved in cultivation, while businessmen and rice millers reap the benefits.

As mentioned before, lack of sufficient and timely supply of water for irrigation is also a major constraint in the agricultural development of Galsi-I and Galsi-II CD Blocks. Besides, lack of law and order in village society has also emerged as a major concern for the farmers. Incidence of thieving of crops from the field and crop damage caused by cattle are also prevalent in many villages.

Table-7.9

Major Problems in Agriculture Perceived by Farmer

S.No.	Major problems in agriculture	Percentage of farmer households
1.	Low yield rate	18.89
2.	Low price of paddy	81.11
3.	Farmers do not get the actual price of paddy, but businessman/miller make profit	51.94
4.	Less amount of land	10.56
5.	Small size Landholdings	7.78
6.	Increasing cost of inputs	75.00
7.	Problem of sufficient and timely water supply for irrigation	42.78
8.	law and order problem	38.61

Source: Field Survey, 2016-2018.

Farmers reported that there is a lack of unity and social cohesiveness among the farmers in villages. No decision related to agriculture is taken unanimously by the farmers at the village level. They therefore urge for an active pro-farmer political farmers' organization. They opined that traders, businessmen and labourers have strong organizations to safeguard their interests, but farmers do not. When we talked about the farmers' outfits of major political parties, farmers argued that most of these sub-organizations either have no power or are inactive at present. These organizations always calculate political gain before taking any action instead of properly addressing their issues. It is also opined that during the Communist period in West Bengal, agricultural labourers and *bargadars* enjoyed political power, while small and marginal owner-farmers got nothing, and at present farmers of all categories are divided from the political point of view. Therefore, they are politically marginalized now.

Agriculture as a Profession for the Next Generation

Due to various causes, farmers do not like agriculture as their main livelihood option. As per the Situation Assessment Survey of Farmer Households, 2003, about 59.4 percent farmer households in India liked farming as a profession, while in West Bengal, only 53.8 percent farmer households liked it. However, no data on the liking of agriculture is available from the reports on the Situation Assessment Survey of Agricultural Households, 2013. As per the Situation Assessment Survey of Farmer Households, 2003, 35.4 percent farmer households in West Bengal did not like agriculture as it is not profitable. Only 1.8 percent farmer households reported this profession lacks social status. About 4.7 percent farmer households opined they did not like agriculture as it is risky.

However, from this study, it is observed that only 25 percent farmer households like agriculture as a profession for their sons or grandsons. In the context of West Bengal, Ray and Ghosh (2007, p.2534) opined, "...today it will be difficult to meet many farmers in West Bengal who are happy to cultivate the crop". About 63.1 percent farmer households opined they do not wish their next generation to be farmers or to be engaged in it. Preferring agriculture as a profession for their next generation is higher among the landless lessees and landless sharecroppers as compared to the other farmer household categories (Table-7.10).

Table-7.10
Prefereing Agriculture as a Profession of the Next Generation

Farmer household categories	Like	Do not like	No response	Do not arise	Don't think about it	Total
Landless lessee	53.3	40.0	6.7	0	0	100
Landless sharecropper	30.0	50.0	5.0	10.0	5.0	100
Marginal	23.4	65.4	2.7	6.4	2.1	100
Small	22.4	64.5	5.3	5.3	2.6	100
Medium	26.2	59.5	2.4	9.5	2.4	100
Big	21.1	73.7	0	5.3	0	100
All	25.0	63.1	3.3	6.4	2.2	100

Source: Field Survey, 2016-2018.

The major causes due to which farmer households would not like to see agriculture as a profession of their son(s) or grandson(s) have been provided in Table-7.11. Unprofitability is the primary cause of disliking agriculture. About 35.7 percent farmer households opined that agriculture is a very labourious activity. Nearly the same percentage of farmer households opined their sons/grandsons are educated, and therefore they wish them to take up other non-farm livelihood option for them. About 24.7 percent farmer households do not want to see their next generation in agriculture, because they believe agriculture can not provide a better life. Declining profitability and declining agricultural land per farmer household, caused by fragmentation of land in the milieu of ever-increasing household expenditure and growing consumerism are the major causes for such a perception.

However, big farmer households who dislike agriculture as a profession of the next generation, rarely consider agriculture as a labourious activity. Interestingly, a percentage of landless lessee households also think so. It is also observed that considerable proportion of farmer households who possess more land think agriculture is insufficient to lead a better life, whereas no one among the landless lessee and landless sharecropper households, and a small percentage of marginal farmer households said they dislike agriculture as their family needs cannot be fulfilled by agriculture. The difference in behavioural space, expectations and household expenditure may be the answer for such contrasting responses.

Table-7.11
Reasons for not Preferring Agriculture as a Profession of Next Generation

Farmer households categories	Agriculture is very labourious and hard activity	Agriculture is not so profitable/ unprofitable activity	Return is insufficient to lead better life	Son/ grandson is educated
Landless lessee	16.7	83.3	0	16.7
Landless sharecropper	60	60	0	40
Marginal	29.3	60.2	18.7	33.3
Small	44.9	53.1	30.6	28.6
Medium	56	68	56	48
Big	14.3	14.3	28.6	57.1
All	35.7	57.3	24.7	35.2

Source: Field Survey, 2016-2018.

In Search of New Sources of Income

Present and past sources of income of the surveyed farmer households have been presented in Table-7.12. From this table, it is evident that most of the farmer households used to depend only on agriculture as a source of income before 1995. However, dependence on agriculture as a sole source of income has decreased over time. At present, about 64 percent farmer households have more than one source of income. Private company job is the new *avatar* among the income-generating sectors for the rural people. No one from the surveyed farmer households was engaged in this sector before 1995, but at present about 5.83 percent households earn income from this sector. Industrial labour, small business, non-farm labour, and different informal activities are also major sources of income for farmer households at present. The percentage of farmer households having government jobs has also decreased. From the economic point of view, increasing share of income from the non-farm sector replacing the farm sector is good for development of a society. Diversified sources of income help the farmer households to fight against sudden odds in any one of their income sources.

However, from the study it has been observed that such engagement in the non-farm sector, in many cases, is not willful, but instead is a forced choice. Some of the non-farm sources of income are riskier than agriculture, hazardous for health, less secure and more exploitative. For instance, working as contractual labourers for bending hot iron rods in an iron and steel industry is very risky. A worker in a rice mill needs to stay for a long time in a dusty environment, which is hazardous for his health. Informal workers in many shops are paid very little. In many cases, members of farmer households need to out-migrate which greatly affects their family as well as social lives.

Table-7.12
Past and Present Sources of Income for Farmer Households

Sources of income	Past (Before 1995) farmer households		Present farmer households	
	in number	in percent	in number	in percent
Agriculture	215	59.72	130	36.11
Agricultural labour	23	6.39	-	-

Agriculture + Agricultural labour	46	12.78	41	11.39
Agriculture + Industrial worker	6	1.67	29	8.06
Agriculture + Other Labours	4	1.11	16	4.44
Agriculture Small Business	43	11.94	65	18.06
Agriculture+ Govt. Service	19	5.28	18	5.00
Agriculture+ Private company's job	-	-	21	5.83
Agriculture+ Other activities	4	1.11	40	11.11
Total	**360.00**	**100.00**	**360.00**	**100.00**

Source: Field Survey, 2016-2018.

We have also enquired on the migration of the members of farmer households (Table-7.13). About 29.7 percent farmer households in the study area have at least one member who migrates for work. The incidence of migration for work declined gradually from landless sharecroppers to big farmer households. Migration among the landless lessee, landless sharecropper, and marginal households is very high compared to the other categories of farmer households. Though some well-educated members willfully out-migrated from their village after getting a job outside, instances of migration for meeting the growing household expenditure is very common. Male members among the landless lessees, landless sharecroppers and marginal farmers migrate for non-farm, informal, daily wage based activities. There are many people from farmer households in every village who engage in agriculture during the day time or during a particular crop season, and migrate at the end of the day or during the off-season for work outside. Another feature of forced migration is that many people have now started to engage in work for which they would not migrate before, fearing loss of prestige in society.

Table-7.13
Farmer Households having Out-migrant Member

Farmer household categories	Farmer households with migrated member(s)	
	Number	Percent
Landless lessee	7	46.7
Landless sharecropper	10	50.0
Marginal	65	34.6
Small	12	15.8
Medium	9	21.4
Big	4	21.1
All	107	29.7

Source: Field survey, 2016-2018.

In this way, agriculture is becoming a part-time economic activity for farmer households. A similar observation was made by Mkhize (2013, p.10) related to the agricultural change in India. According to him, "There is therefore a marked tendency of agriculture to move to a productive and modern model of part-time farming" and "structural transformation in the form of a decline in agricultural employment and in favour of non-agricultural employment is happening in India". Multi-occupational dependancy has become a dominant feature of the present-day Indian countryside (Yadu & Satheesha, 2016, p.22). However, Mkhize (2013) expressed his concern regarding such a structural transformation of the Indian economy. His concerns also perfectly fit with our observations. In Mkhize's words (2013, p.10), "...the new form of structural transformation is a stunted one because it primarily generates employment that is informal and/or insecure, without the benefits of health and unemployment insurance and pensions".

Conclusion

From the analysis, it is observed that most of the landless

lessees and landless sharecroppers are SC, and these two categories of farmer households are more impoverished compared to the other farmer household categories. It may also be said that farming society is on the verge of a tremendous change in the near future. Growing consumerism, modernity, asset drift, increasing dependence on the non-farm sector, and forced migration of people from farmer households are the characteristics of farmer society these days. All the consequences of these changes in farmer society are not uniformly good. Moreover, some changes like an increase in LPG use, number of toilets etc., are the result of implementation of government programmes and are definitely good for them.

Besides the pressure of population growth on land and fragmentation of landholdings, spread of modernity and consumerism into the farmer society are creating immense financial pressure on farmer households at a time of job crisis and agrarian distress. To get rid of this pressure, members of farmer households are being compelled to take up under-paid, less secure, and informal jobs/economic activities. Declining savings, disproportionately rapid growth in expenditure as compared to income growth and forced migration can never be a sign of sustainable economic development, even if the standard of living improves.

As the study area has a high percentage of agricultural labourers, development of agriculture as well as the creation of ample job opportunities in the non-farm sector is essential. Otherwise, the distress of farmer society would be aggravated. In the current phase of economic slowdown in India and in the period of jobless growth, some sort of distress has already increased in the study area leading to increasing short-term tenancy. The issues around short-term tenancy has been addressed in the next chapter.

8 Short-Term Tenancy and Further Distress

Introduction

The previous chapter highlights that most farmers as well as new generation do not like agriculture as a profession. Growing household expenditure, mostly caused by recent behavioural changes in the farmer society also demands non-farm employment opportunities. However, in the development of the rural economy in India agriculture has played a vital role till date. In recent years this sector has been experiencing two contradictory characteristics. One is that agriculture is becoming an unprofitable economic activity (Raghavan, 2008; Narayanamoorthy, 2013; Tripathi, 2013). The other is tenancy cultivation, more specifically fixed rent tenancy, which is increasing day by day though it is not permissible in many states (Haque, 2012; Mani, 2016). In the post economic reform period, cost of food production has increased but food prices remain low (Nair & Eapen, 2015). As a result, a large proportion of the people engaged in the farm sector are facing serious erosion in their income and welfare. Profitability issues have been already discussed in Chapter 4.

From the study by Ranganathan (2015, p.57-58), it has been observed that in India, proportion of households leasing-in land was 12.76 percent in 2002-03, which increased to 16.42 percent in 2012-13. West Bengal experienced higher growth rate in comparison to the national level. In this state, percentage of farmer households leasing-in land was 19.11 percent in 2002-03, which increased to 37.18 percent in 2012-

13. West Bengal ranked first in proportion of farmer households leasing-in land in 2012-13. Thus, it is clear from the data that not only is the number of participants in the land lease market in West Bengal increasing, but this is happening at a very high rate in comparison to other states. Besides farmer householdss, NSSO data from different rounds reveal that percentage of leased-in area has also increased significantly since 2002-03. The percentage distribution of leased-in area in India has increased from 6.60 percent in 2002-03 to 10.41 percent in 2012-13. A similar trend is also visible in case of West Bengal. During the same period, this state experienced growth of 5.31 percent (9.42 percent in 2002-03 increased to14.73 percent in 2012-13).

At present, like in many other states in India, fixed rent tenancy for a short period is experiencing huge growth in West Bengal, although short-term fixed rent tenancy (FRT) is not new here. Before independence, fixed rent tenancy was present in Bankura, Mednipur, Murshidabad and Nadia districts (Bhaumik, 1993). In 1982, 14.75 percent of leased-in area was under fixed rent tenancy, whereas 55.83 percent of the leased-in area was under share tenancy (Bhaumik, 1993, p.61). Later on the FRT system increased instead of share tenancy. It is reported by Mani (2016) that in 2012-13, about 52 percent of the leased-in area was under fixed rent tenancy in West Bengal. Under FRT, fixed amount of crop or fixed amount of money is paid by tenants as rent. FRT began as a fixed crop contract which has since been changed into a fixed money contract. A similar situation is observed in Purba Bardhaman district as well as in the study area, where the dominant practice is short-term fixed money tenancy. Very recently, suicide of paddy cultivators in Purba Bardhaman district (Anandabazar Patrika, 2015, 2017a, 2017b) has also started being reported in newspapers. In all these news items, it has been observed that farmers committed suicide due to crop failure and all of them held leased-in land contracts under short-term FRT.

In this context, this chapter tries to investigate: a) why,

in spite of tremendous risk involved, is informal short-term fixed rent tenancy increasing in the main rice growing region of West Bengal; b) who are participating in the process; c) what are the specificities of this tenancy system; and d) what are the impacts of this practice on the lessees and on agriculture. In addition, history of tenancy relation and tenancy systems in practice, have been discussed in brief for understanding tenancy context in West Bengal.

There are two parties in land-lease market: lessors[21] and lessees[22]. Lessor households have been classified into four size groups: a) marginal with ownership holding of less than 1 hectare; b) small with ownership holding of 1-2 hectares; c) medium with ownership holding of 2-4 hectares; and d) big with ownership holding of greater than 4 hectares, following the classification of farmers by the Directorate of Agriculture, Govt. of West Bengal. The lessee households have been classified into six categories. These are: landless labour, agricultural labour cum marginal farmer, marginal farmer, small farmer, medium farmer and big farmer.

This chapter is based on primary data collected from field survey. In the study area, out of 360 farmer households surveyed, 66 farmer households of lessee cultivators and 35 lessors households who used to take part in short-term land leasing were detected and surveyed separately to understand this tenancy system. For in-depth analysis, key informant survey and focus group discussion were also conducted. Percentage analysis method has been used for interpretation of the collected data.

A Brief History of Tenancy System in West Bengal

The tenancy system in India and Bengal is as old as the country itself. In the Mughal period, the peasants continued

21. A person who owns agricultural land and has leased out that land to a lessee cultivator, under mutually agreed terms and conditions (NITI Aayog, 2016).
22. A person who leases-in the agricultural land against a consideration in cash or kind or a share of produce payable to the landowner-lessor as per the lease agreement (NITI Aayog, 2016).

to enjoy customary rights over the land occupied by them. In Mughal period, there were mainly two types of cultivators: a) *khudkasht ryots*, or residential cultivators, b) *paikasht ryots*, or non-residential cultivators. One class of *paikasht ryots* would supply a part of inputs and pay land revenue to the superior at a pre-determined rate either in cash or in produce. Another class of *paikasht* was *bataidar* or sharecroppers (Bandyopadhyay, 2013, p.39).

The British government introduced the first tenancy legislation, the 'Rent Act of 1859', aiming at strengthening the position of peasants on their land (Swamy, 2010, p.9). This act classified *ryots* into three categories (Bhaumik, 1993, p.28): a) *mukuraridas* or *istamraridas* who cultivated land uninterrupted at fixed rate for more than twenty years; b) occupancy ryots who cultivated land uninterrupted for a period of twelve years; and c) non-occupancy ryots who could not prove twelve years' uninterrupted possession over land. *Bargadars* (registered tenants) were included in this third category. However, colonial land policy helped to spread the process of sub-infeudation. During this time, sharecropping spread mainly due to the *jotedari* system, commercialization of agriculture, de-industrialization, small landholdings, colonial modernity and growing job opportunities for the educated middle class (Mukhopadhyay, 1992). However, sharecroppers faced exploitation by landowners and had no security of tenure.

In 1936, the All India Kisan Sabha was formed to safeguard the interests of sharecroppers. This organisation initiated the *Tebhaga* movement of Bengal. Fought in 1946-47, this was a significant one focusing on the two-thirds crop demand of the sharecroppers. However, the situation of the sharecroppers did not change much for a few years even after independence. The first tenancy legislation in Independent India was the 'West Bengal Bargadar Act of 1950', which can be considered as a major outcome of the *Tebhaga* movement. Though several tenancy legislations were passed between 1950 and 1977, there was lack of political will for effective implementation. In 1978, the Left Front government started '*Operation Barga*' to record names of *bargadars*. The government also took steps so that *bargadars* could avail institutional credit facilities. This movement achieved great success and became a milestone in the history of tenancy reform in India. At present, there are very few states like West Bengal where so many tenants have been recorded by the government.

Upto the 9th Five Year Plan, almost all states passed restrictive laws regarding tenancy. Following this trend, West Bengal had also banned fixed rent tenancies though sharecropping was permissible (Hanstad, 2005, p.15). The 10th Five Year Plan acknowledged that the ban on tenancy created concealed tenancy and thus provided no protection to the tenants (Haque, 2012, p.5). To some extent this is true for West Bengal also, as there are many tenants who are engaged in fixed rent tenancy though it is prohibited. However, after *Operation Barga*, no such steps have been taken towards tenancy reform in West Bengal.

Tenancy System in Practice

Two main categories of tenancy-sharecropping and fixed rent tenancy-have been observed in the study area along with many other sub-types under those broad categories. Sharecropping is mainly practiced during *kharif* season. In this system the landowner allows the tenant to use the land against a share of crops produced on their land. Generally, a registered tenant (locally called *bargadar*) gets 75 percent share of crop if he provides non-labour inputs (termed as sharecropping without cost sharing) and gets 50 percent share of crop produce if landowner provides the non-labour inputs (termed as sharecropping with cost sharing). However, there are also several unrecorded sharecroppers as observed in the study area. In unrecorded sharecropping, different crop share ratios exist depending on the personal relation between tenants and landowner, and/or different types of cost sharing. Sometimes, the crop sharing ratio also depends on quality and location of the land. In some villages where irrigation facility is quite good, unrecorded sharecroppers get 50 percent crop share if they bear half the cost of ploughing, fertilizer and seeds. Unrecorded sharecroppers get 2/3rd of the crop produce if landowners do not provide anything except land. This type of crop sharing is known as the *tebhage* system. There is also another type of crop sharing, which is known as the *krishani* system. It is a special type of share cropping in which landowners provide land, seed, fertilizer and other costs except labour and irrigation costs, and get 2/3rd to 3/4th share of total produce. Generally, resource-poor scheduled tribe people take part in this system.

Under the fixed rent tenancy (FRT) system, land owners lease out land to get a fixed amount of crop or money as rent for the land. In many areas, lessors get apart or total of stipulated rent in advance. Rent in terms of money is more or less equivalent to the price of 'fixed' crop amount. Though fixed rent tenancy is common during *boro* season, it is also observed during *kharif* season.

In another trend, many sharecroppers who have been cultivating land for more than 10-15 years under the same landowner are now paying a fixed amount of crop rather than the share of the crop. We have identified 56 sharecroppers out of which 44 tenants recently opted to pay a fixed amount of crop, which is more or less equivalent to 1/4th of average production. The landowners of these sharecroppers are mostly (76.79 percent) engaged in non-farm sectors and live in the nearby town. Because of their absence from the village, it becomes difficult for them to enquire about the actual crop production in each and every season. Therefore, they also prefer to 'fix' the rent in accordance to the crop share they got on an average during the previous years. In this particular case of fixed crop tenancy, rent is comparatively lower than the rent in common short-term FRT because of their personal relationship with the land owner as sharecroppers over a long time. However, these sharecroppers turned fixed rent tenants are not included under the short-term fixed rent tenancy system.

Short-Term Fixed Rent Tenancy and Land Lease Market

Generally, households with an abundance of labours but insufficient land to utilize, can lease-in land from those who have land in abundance but lack the necessary manpower to utilize it fully (Goswami & Bezbaruah, 2013, p.65). According to Haque (2012, p.15), landless, marginal and small farmers constitute about 98 percent of the lessees in West Bengal. However, this figure does not differentiate between sharecroppers and short-term lessees. Thomas and Indira Devi (2015, p.145) in their study in Kerala showed that lessors are highly educated and dependent on non-farm activities, whereas the lessees are less educated and dependent on agriculture. Ahmed (2011, p. 20), in the context of Bangladesh, also observed that lessees are less educated and poorer than the lessors. However, she opined that "having more opportunities of stable employment strengthens the tenant's tolerance to risk and encourage them to choose fixed rent contract" (2011, p.20). Using NSSO data of the 59th round, Sharma (2007) in his study showed that in India, almost 90 percent of the lessees are from landless and marginal farmer households. Sharma has also found that most of the lessees come from landless, marginal, small and medium categories of farmer households, and are from 'other backward class' (OBC), followed by 'other castes' and 'scheduled castes' (SC). On the contrary, though the highest proportions of lessors are marginal farmers, the tendency for leasing-out land is highest among the big farmers and among those belonging to 'other castes'.

Among the surveyed households in the study area we detected 18.33 percent of households are engaged in short-term fixed rent leasing-in, whom we address as lessee households. The landless labours and marginal farmers emerge as significant participants in leasing-in. These categories together contribute 78.82 percent of households participating in short-term leasing-in (Table-8.1). Medium and big farmers constitute only 3.03 percent and 1.51 percent respectively. This figure is quite similar to the observations of studies (Sharma, 2007; Haque, 2012; Vijay & Sreenivasulu, 2013) from other parts of India.

It is very interesting to note that about 20 percent of the lessees are just landless labours. Generally, resource-poor landless labour households used to lease-in land in case of sharecropping in the study area. However, in FRT lessees have to invest total input cost, for which they need capital investment even on a small scale. As a result the proportion of landless households participating in the short-term FRT is quite less. In this short-term FRT, another important observation is the caste-wise distribution of lessees (Table-8.1). The lessee households are drawn in significant proportion from the SC category (47 percent) followed by OBC (30.3 percent). SCs are also highly represented in 'landless agricultural labours', 'agricultural labours cum marginal' and 'marginal household' categories. Only 3.03 percent of the lessee households are from ST category, because of their low level of concentration in the villages of the study area.

Table-8.1
Size Category and Caste-wise Distribution of Households Leasing-in

Lessee categories	Proportion of households leasing-in	Caste of the lessee			
		General caste	SC	ST	OBC
Landless labours	19.70	7.69	69.23	-	23.08
Agricultural labour cum marginal farmer	16.67	-	63.64	9.09	27.27
Marginal farmer	42.42	17.86	46.43	3.57	32.14
Small farmer	16.67	54.54	9.09	-	36.36
Medium farmer	3.03	50.00	50.00	-	-
Big farmer	1.51	-	-	-	100.00
All	100	19.70	46.97	3.03	30.30

Source: Field Survey, 2016-2018.

Among the lessors, about 48 percent (Table-8.2) are from marginal farmer householdss, followed by small (28.57 percent), medium (14.29 percent) and big farmer households (8.57 percent). Thus, small and marginal farm households together constitute about 77 percent of the lessors. The detailed investigation and analysis of the lessors shows that among the lessor households majority belong to the general caste and OBC (Table-8.2). Thus, small and marginal farmer households dominate in this land lease market. The total percentage of small and marginal farmer households participating in leasing-in and leasing-out under this short-term FRT is more or less same (around 78 percent in both cases). Medium and big farmer households tend to lease-out their land more than lease-in.

Table-8.2
Size Category and Caste wise Distribution of Households Leasing-out

Lessor categories	Proportion of households leasing-out	Caste of lessors			
		General caste	SC	ST	OBC
Marginal farmer	48.57	47.06	17.64	-	35.29
Small farmer	28.57	70.00	20.00	-	10.00
Medium farmer	14.29	60.00	-	-	40.00
Big farmer	8.57	66.67	-	-	33.33
All	100.00	57.14	14.29	-	28.57

Source: Field Survey, 2016-2018.

In terms of education, as shown in Fig. 8.1, lessees have low level of education. Of the lessees, 62 percent have primary level education, 31 percent have secondary to higher secondary level education and few are graduates. As opposed to the lessees, a majority of the lessors have higher level of education. As many as 57 percent lessors have secondary to higher secondary level education and more than 15 percent are graduates. Only 6 percent of the lessors have just primary level education. Different studies on agricultural tenancy (Ahmed, 2011; Kar & Banerjee, 2010; Vijay, 2012) have also

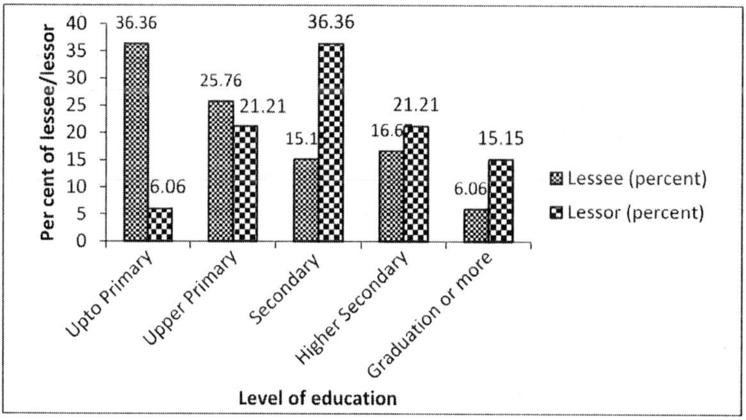

Fig. 8.1: Educational Status of Lessees and Lessors

Source: Field Survey, 2016-2018.

shown that less educated families generally lease-in land.

Considering the class, caste and education together in the short-term fixed rent land lease market, it is found that most of the lessees are marginal farmers followed by landless labours and small farmers, and they mostly belong to SC category. Surprisingly, most of the lessors are also marginal farmers, but they belong to the general caste, followed by OBCs. Participation of ST households in this tenancy system is very negligible because of their poor percentage of presence in the area. The level of education of lessees is much lower in comparison to the lessors. Thus the short term lease-in is more common among backward caste people with low level of education, supported by family labour and reduced input cost of production.

The analysis of all sources of income of individual households led us to identify the dominant groups acting as lessor and lessee. Among the lessors, about 33 percent have no other source of income except agriculture and around 67 percent are engaged in the non-farm sector. Interestingly 29 percent of the marginal farmer lessors and 37.5 percent of the small farmer lessors have no other source of income except agriculture. This observation indicates that even if people are solely dependent on agriculture, small and marginal farmers do not take the risk of cultivating their own agricultural land (discussed later) and they also opt for leasing out.

Among the lessee households, more than 24 percent have no other source of income, 27 percent are agricultural labours, 3 percent are engaged in agriculture-allied activities and the rest of the households are engaged in non-farm sector activities. Around 32 percent of the marginal farmer lessee households and 54 percent of the small farmer lessee households have no other source of income except agriculture. Medium and big farmer lessee households sometimes have additional income from other petty business, from which they use the capital to invest in land leasing-in. Thus, considering all possible sources of income, we can identify four categories of households that are participating in leasing-in land: a) Landless agricultural labour and agricultural labour cum marginal farmer households; b) Marginal and small farm households who have no other source of income except agriculture; c) Households having financial support from non-farm employment of one or two family members working outside. They may be from almost all categories except medium and big farmers; and d) Medium and big farmer households who have sufficient capital to invest to enhance their income from the agricultural sector.

The Reasons of Increasing Short-Term Fixed Rent Tenancy in Agriculture

There is an enormous body of literature on fixed rent tenancy, where the causes for land leasing have been analyzed. Generally it is observed that the disparity in the allocation of land and workers is the main reason of increasing short-term FRT. In agriculturally prosperous areas with canal irrigation, due to expansion of non-farm sector and growing demand for land, rich peasants are becoming non-cultivating and gaining income by leasing-out land in spite of selling their 'asset' (Vijay & Sreenivasulu, 2013, p.49). Besides economic factors, there are other factors as well.

It is important to know the causes behind growth of FRT over sharecropping. Rudra (2014, p.22) opined that due to the advancement of technology, production has increased and consequently the tenancy system has changed from sharecropping to fixed rent tenancy for crops like summer paddy. This observation is partly true here as well. Lessees do prefer FRT considering the fact that if they can produce more crops using more fertilizer and other inputs, they will get more returns compared to sharecropping. However, Ahmed (2011, p.8) in his study on tenancy contracts in Bangladesh opined that as there is risk of crop loss due to low productivity, lessors always prefer fixed rent tenancy over sharecropping, but lessees generally prefer share tenancy due to risk aversion. In the report of NITI Aayog (2016), it has been pointed out that restrictions and sometimes ban on FRT laws in different states of India are also responsible for the growth of informal short-term FRT. In the study area, the following reasons stand out as important factors of increased leasing-in and leasing-out in short-term FRT.

Leasing-Out

In Galsi-I and II CD Blocks of Purba Bardhaman District, with the canal water provided by DVC and groundwater tapped by deep tube wells, farmers used to produce *boro* paddy, which is a highly profitable crop in comparison to *aus* and *aman*. However, in the recent past, the DVC has failed to provide adequate irrigation water in time, and groundwater level has also declined, limiting the supply of groundwater. As a result of unreliable supply of water, *boro* paddy cultivation has become more risky. Besides, input cost has also increased rapidly compared to the increase in the price of paddy. Roy and Ghosh (2007) in their study on *boro* paddy cultivation in West Bengal showed that after 1980-81, net income from *boro* paddy cultivation has been in gradual decline. Contrary to this, cost of cultivation has increased, mainly due to increase in costs of fertilizer, labour and rented agricultural implements. As a result, farmer households which had been cultivating *boro* paddy for a long time are losing interest in it.

Moreover, as social upgradation is taking place among rural people, it is not possible for them to manage the growing family expenditure with the declining farm income. As a result of the aforesaid situation, many farmer households are not cultivating their agricultural land fully by themselves, and some of them search for other types of jobs from where income is greater than farm income. Many farmer households cultivate their own land only for their food consumption requirement during the *aman* season. During *boro* season, instead of taking the risk of farming by themselves, they lease-out. Most of the lessor households, during interviews, stated that they do not lease-out land during *aman* season as this type of paddy cultivation needs less amount of fertilizer and less irrigation water, but assured returns compared to *boro* paddy. Moreover, during harvesting period (first half of November), there is low risk of extreme weather, minimizing the chances of crop loss. In contrast, *boro* paddy cultivation needs higher amount of fertilizer, more irrigation water and often is badly affected by hail-storms, unusual rainfall and storms caused by nor'wester during the harvesting period.

The high yielding varieties (HYV) of seeds such as MTU-7029 and MTU-1010 are largely used for *boro* paddy, which often get germinated during harvest even after low rainfall in the hot summer season. Considering all the risk factors involved in the *boro* crop, the land owners often lease-out their land. They also admitted that earlier, they used to take the risk because of higher price compared to production cost. In the last 20-25 years, production cost as well as the price of consumer goods has increased sharply as compared to the rise in the price of paddy. That is why they have lost interest in *boro* paddy cultivation by themselves. Overall, agriculture is not a profitable activity now. Though the profit margin has declined, farmers still would like to retain their land as they have an emotional attachment with the land and consider it as a valuable asset that can give regular returns in terms of rent from leasing-out.

Labour crisis in terms of both supply and wage is another cause of leasing-out. During the 1980s and early 1990s, agricultural workers used to come from all over West Bengal and from eastern and southern part of Bihar seeking employment in Bardhaman during sowing and harvesting season (Rogaly, 1999, p.360). Many older farmers opined that due to socio-economic upliftment of labourers in the backward areas, advancement of technology and declining area under *boro* cultivation, seasonal migration of agricultural workers has almost stopped. Now only women labours commute locally from labour surplus villages to labour deficit villages for higher wages.

Though this region has abundant labours, most of them out-migrate to other urban-industrial areas both inside and outside the state to work as contractual labour. The young, moderately educated male labour class is no more interested to work in the agricultural sector. The cultural change has impacted them to consider other non-farm informal work as better in comparison to agricultural work. They get more working days and sometime higher wages than they would receive as agricultural labours. During interviews, many small and marginal farmers blamed the medium and big farmers for the labour crisis, as they pay advance money to the agricultural labours forcing them to work at their land first. Many lessors also blamed MGNREGS for higher wages and degradation of work ethics. At present, most agricultural labours prefer to work on contract basis rather as daily labours. This is due to fixed quantum of work, they can earn two wages in a single day. Due to the agricultural labour crisis, remaining agricultural labours are now dominating the labour market. There is also lack of family labour among the better-off farmer households, which forces farmers to lease-out their land for short-term fixed rent tenancy. About 20 percent lessors admitted that the labour crisis is the main cause for leasing-out.

In some villages such as Sanota of Galsi-II and Jagulipara of Galsi-I CD Blocks, tenancy rate is very high due to high demand for land leasing-in. In these villages, many farmer households responded that they have started to lease-out recently because of low profit margin on one hand, and higher rent from land on the other hand. In this situation, leasing-out for short-term FRT is more profitable as there is no risk sharing; rather getting assured returns. About 17 percent lessors opined that short-term FRT is more profitable than cultivation by themselves.

Low risk-bearing capacity is another cause for leasing-out. Marginal farmers with scanty landholdings produce *aman* paddy for subsistence, but do not take the risk of cultivating *boro* paddy in the years they do not get a profitable price from *aman* paddy. Most of these farmer households neither have bullocks nor tractors for ploughing and transport. All these households had sufficient farm implements earlier. Now, with the use of tractor in place of cart and animal-drawn plough, and chemical fertilizer in place of bio-fertilizer, they have lost their self-sufficient practices and become dependent on hired implements. These resource-poor farmer households are failing to make a profit cultivating their land with hired implements and hired labour. Therefore, they are leasing-out for short-term FRT.

Another cause of growing short-term fixed rent tenancy is the fear of *barga* record among the lessors. West Bengal experienced a successful implementation of tenancy reform through *Operation Barga*. Memory of that time still exists in the mind of farmers. In West Bengal, tenancy law does not provide ownership right to the sharecropper, but it gives permanent hereditary right to cultivate *barga* land. Therefore, most of the lessors are cautious in leasing-out for longer term. They prefer to lease-out for one crop or for one year only. Sharecropping needs continued monitoring by the owner during the harvesting period to ensure correct returns, which becomes difficult for owners due to time constraints for their other work, thus increasing the chances of fixed rent contracts replacing sharecropping.

Leasing-in

Leasing-in is also increasing for various reasons. In many studies, it has been argued that lack of alternative job opportunity is the main cause for leasing-in by the poor tenants (Vijay & Sreenivasulu 2013; Mohanakumar, 2014). Kar & Banerjee (2010, p.132) point out that a large section of wage earners lease-in land to arrange at least part of the rice required for consumption (which they term supplementary farming). Peasants, who have sufficient labour but do not have access to land, lease-in land. Agricultural labour and poor small and marginal farmers are interested in land leasing as it provides them assured employment over the year (Bhaumik, 1993) and they can use the surplus labour on leased land, if they fail to find gainful employment opportunities (Vijay & Sreenivasulu, 2013, p.47) in other sectors. Mohanakumar (2014) in his study on tenancy relations in Travancore in South Kerala blamed the loss-making agricultural system for growing tenancy. To him, loss in employment days in the crop production sector 'forced' wage labour and pure tenants to lease-in land.

There is tremendous socio-cultural rural transformation happening in India during the neo-liberal regime. Profit margins in agriculture have declined. Job opportunities have also not been created at a desired level. However, due to social upgradation of households and inflation, expenditure has increased to a remarkably higher level. Lessee households lease-in land for extra income to supplement the additional expenditure. Amount of agricultural land per household is also decreasing due to division of land through breaking of joint families. Agricultural labour cum marginal farmers lease-in land as they solely depend on the agricultural sector. Generally, being less educated and/or associated with this sector for a long time, they prefer not to work outside the village. They think that tenants have more social prestige than the agricultural labour. Sufficient household labour also motivates them to lease-in land, as it apparently lowers the production cost.

Marginal and small farmers, who have no other source of income except agriculture, lease-in land mainly to cope with the growing household expenditure. Having failed to find suitable non-farm work, they invest their labour in cultivating leased-in land along with their own land. Many of them admitted that they are being pressurized by the demands of consumerist society. Because of this increasing demand they are trying to generate more income through short-term FRT cultivation. However, about 53 percent households under this category either have shallow/deep tubewells or have bullock-drawn plough/tractor. These households also choose to lease-in land so that they can fully utilize their investment on farm implements.

Some of the landless labours, agricultural labour cum marginal farmers, marginal and small farmer households of which members are also engaged in non-farm activities, have been leasing-in land. In this case, two factors have been observed. First, very poor farmer households invest part of their income from the non-farm sector in lease cultivation to ensure they have enough rice for consumption throughout the year. Second, having failed to invest in non-farm income in any other business or productive sector, many marginal and small farmer households invest in lease cultivation for additional profit making. They generally lease-in land adjacent to their own land to minimize the input cost and supervision time.

Very few medium and big farmers who have tractors and submersible or shallow tubewells also lease-in land. They take advantage of technological advancements and better financial condition to get the scope of economies of scale. Whereas poor farmers lease-in land for improving their earnings, "some medium and large farmers also lease-in land for full utilization of their fixed resources such as tractor, irrigation pump sets, etc." (Haque, 2012, p.8).

Thus there are two main causes for leasing in: a) economic distress and b) profit making. Farmer households, who have sufficient money or farm implements, are mainly driven by the second reason. During our field survey in Ramgopalpur village, we interviewed a small farmer who has a submersible pump. He plays a dual role in the lease market. He leases-out land in the years when the irrigation department provides canal water for *boro* cultivation, as in these years demand for land among the poor farmers remains high, resulting in higher rent. Contrary to this, he leases-in land in years when the canal remains dry, resulting in low demand for leasing-in, which in turn results in low rent for the leased land. It makes lease cultivation profitable for him. He informed that many medium and big farmers of the surrounding villages also follow the same strategy.

Specificities of Short-Term Fixed Rent Tenancy

In the study area it has been observed that most of the lease agreements are informal. About 92.08 percent of the lease contracts are determined mutually by lessees and lessors on oral basis. In case of 7.92 percent lease contracts, lessors and lessees made a written agreement on a plain paper. Such cases have been found at Jagulipara, Bhirsin and Chaktentul of Galsi-I CD Block.

About two-third of the lessor households lease out their land every year (Fig. 8.2). Few lessors lease-out their land intermittently. In Galsi-I and Galsi-II CD Blocks, canal irrigation water is not available for *boro* paddy cultivation for every year. There is quota of providing canal water for *boro* paddy cultivation. As a result, some of the lessor households

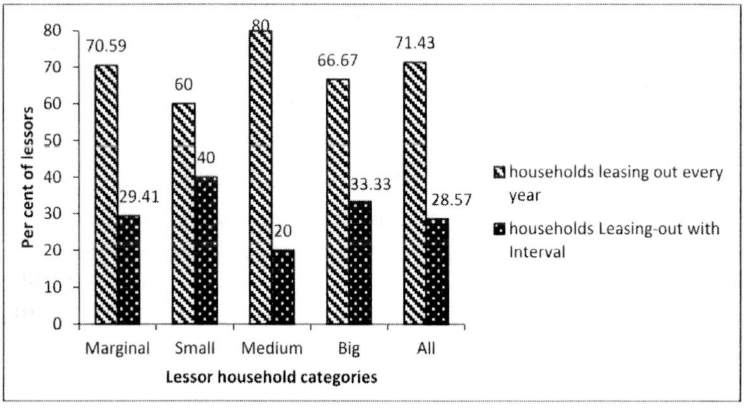

Fig. 8.2: Lessor Households by Frequency of Leasing-out
Source: Field Survey, 2016-2018.

cultivate their land by themselves instead of leasing-out in which year canal irrigation facility is provided for *boro* paddy cultivation.

Similarly, most of the lessee households lease-in land every year to earn sustainable income. About 86.36 percent lessees leased-in every year, of which most of the lessees are small and marginal farmers and landless labours (Fig. 8.3). Therefore, it can be said that this type of leasing is mainly driven by hunger leasing. Most of the medium and big farmer households lease-in land when rental rate allows them to make profit.

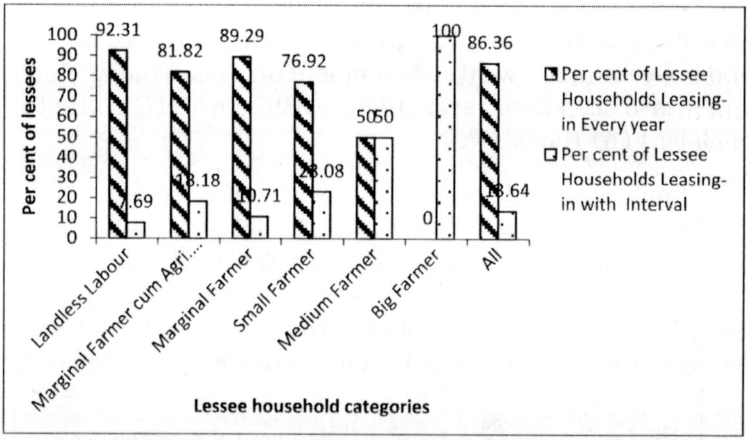

Fig. 8.3: Distribution of Lessee Households by Occurance of Leasing-in

Source: Field Survey, 2016-2018.

In the study area, about 74.26 percent of the lessors and lessees admitted that they are only known to each other (Table-8.3). Lessors lease-out their land to those lessees who promise to give maximum rent. About 23 percent lessors lease out fraction of their agricultural land to the labour household who provide labour at the lessors' land during crucial time.

Table-8.3
Relation between Lessee and Lessors

Nature of relation	Percentage of respondents		
	Lessors	Lessees	All
Work as agricultural labour at lessors' field	22.86	13.64	16.83
Relative	2.86	10.61	7.92
Known only	74.28	74.24	74.26
Friend	-	1.51	0.99
All	100	100	100

Source: Field Survey, 2016-2018.

Lessors generally prefer to lease-out their land only for short time period. From the Fig. 8.4, it is observed that 45.45 percent lessees admitted lease contract remain valid only for one crop season. About 41 percent lessees said lessors lease out their land to them for one years. Only 13.64 percent lessees have succeeded to make lease agreement of more than one year duration.

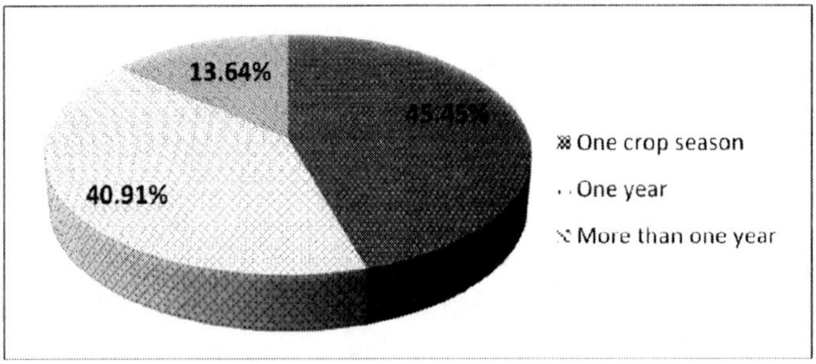

Fig. 8.4: Distribution of Lessee Households by Duration of Lease Contracts

Source: Field Survey, 2016-2018.

As this region had experienced successful implementation of *Operation Bagra*, lessors have the 'fear of land recording'. Therefore, most of the lessors (83.86 percent) do not lease out their land to the same person for long period. Besides, more rent offered by the other lessees also influences lessors to change the lessees. Only 17.14 percent lessor households told that they lease-out land to the same lessees

(Table-8.4). Tendency of changing lessees each time is highest among the big farmers (100 percent) followed by small and medium farmers.

Table-8.4
Change of Lessee by Lessors

Lessor Category	Percentage of Lessor Households		
	lease out land to same lessee	Change lessees each time	Change lessee after few years
Marginal farmer	29.42	35.29	35.29
Small farmer	-	80	20
Medium farmer	20	60	20
Big farmer	-	100	-
All	17.14	57.14	25.72

Source: Field Survey, 2016-2018.

This type of land leasing is mainly practiced during *boro* paddy cultivation. It is found that 77.14 percent lessors lease-out their land only in *boro* season (Table-8.5). Only 20 percent of the lessors lease out for both *boro* and *aman* paddy cultivation. Among the lessors 8.57 percent are non-cultivating peasant households as they do not take part in farming all over the year. About 54 percent lessor households lease out their whole cultivable land. 37.14 percent lessors lease out up to half of their cultivable land. These lessors, as it is told by them during interview, lease-out those land holdings which are far located and/or have lack of irrigation facility. However, in some cases lessors cannot lease-out their total amount of land as all the land holdings are not suitable for *boro* paddy cultivation.

Table-8.5
Extent of Leasing-out

Percent of cultivable land leased-out	Percentage of Lessor households			
	In *aman* season	In *boro* season	In *aman* and *boro* season	Total
Upto 50 percent	2.86	25.71	8.57	37.14

Upto 75 percent	-	5.71	2.86	8.57
Upto 100 percent		45.71	8.5	54.29
Total	**2.86**	**77.14**	**20**	**100**

Source: Field Survey, 2016-2018.

The agreements under fixed rent contract differ from one village to another and even within village. Generally, rent in kind varies from 3 sacks to 6 sacks of 60 kg. per *bigha* with an average rent of 4 sacks during *boro* paddy cultivation in those villages where canal irrigation is available. The cash rent for *boro* paddy varies from Rs. 3200 to 5000 per *bigha* in canal irrigated area. However, rent is relatively low if the land is irrigated by the submersible. For this, rent in kind is 2 to 3 sacks of 60 kg. and rent in cash is Rs.1500 to 3000 per *bigha*. Though average rent in kind or cash is more or less equivalent to 1/4th of the average rate of production, it becomes higher in which year competition for leasing-in prevails. In Ramgopalpur of Galsi-I CD Block and in Sanota, Adrahati of Galsi-II CD Block competition exists for leasing-in land among the poor and rich peasants.

During *aman* season, rent in kind varies between 3 to 4 sacks of 60 kg., whereas rent in cash varies between Rs. 2500 and 3500. In this tenancy, generally rent is paid after harvesting of crop. However, recently lessors are demanding payment of rent at the time of leasing-out. In Kolkol and Jagulipara villages of Galsi-I CD Block, such type of practice is highly observed. Uncertainty over getting the stipulated amount of rent in the years of crop failure instigates lessors to demand payment of rent in advance. Some lessors also blamed that after harvesting the paddy, many lessees falsely claim loss, thus provide less amount of crop or cash as rent, and as a result they are compelled to take the rent amount in advance.

Impact of Short Term Fixed Rent Land Leasing

In traditional economic literature, fixed rent tenancy is viewed to be more efficient than sharecropping (Goswami & Bezbaruah, 2013, p.60). However, fixed rent tenancy places the entire risk of crop failure on tenants (Ahmed, 2011, p.8). In our study it has been observed that almost 42 percent lessee households have experienced losses in short-term FRT cultivation. Most of the lessee households which faced losses are agricultural labour cum marginal farmer (72 percent) followed by landless labour (61 percent). These two groups faced losses because of the lack of required farm implements. Among the landless labour households, 69.23 percent have not a single farm implement, 23 percent have spray machines, and around 31 percent have threshers. Only one household has bullocks and cart. About 45.45 percent of the agricultural labour cum marginal farmer households do not have any farm implements.

Moreover, as this type of tenancy is informal, most of these lessee households are compelled to depend on informal sources of loan with higher interest. Among the lessee households, only 40.90 percent possess a kisan credit card (KCC). Landless labours have no KCC and only 27.27 percent of agricultural labour cum marginal farmer households possess KCC. Therefore, most of the lessee cultivators can not avail formal sector loans at low interest rates. We have traced two Joint Liability Groups (JLGs) which could help informal tenants to get loans from formal sources, but in reality their role is very insignificant. Now, due to higher input cost and loans at higher interest they are generally compelled to sell off their produce instantly after harvesting it, at a lower price to the local trader. About 41 percent lessee households have reported distress sales. However, collected data may not properly reveal the intensity of such loss because most of the farmer households do not calculate cost related to household labour, depreciation, own bullocks and labour, etc. If all of these are included in input cost, percentage of lessee households facing losses would be higher than it appears.

From the Table-8.6 it is observed that only 16.67 percent lessee families have failed to fulfil their lease agreement, though 42.42 percent of them faced losses. Thus, it can be said that lessee families have been compelled to fulfil lease agreements in any way possible. Even if the lessees fail to pay the stipulated rent, they have to pay the dues in the next season. We have asked lessees why they fulfil the lease agreement in spite of incurring loss. At this, they answered that if they failed to pay, their reputation will be tarnished, and as a result next time land-owners will be unwilling to lease-out their land to them. Advance payment in this tenancy system is also increasing day by day. This type of agreement is pushing the landless labour, marginal and small farmer lessees into huge risk.

Table-8.6
Proportion of Loss and Fulfilment of Lease Contract

Categories of Lessee	Proportion of loss in short-term Fixed rent tenancy			Fulfilment of lease-contract (%)	
	Faced loss	Faced no loss	No response	No failure	Failed to fulfil
Landless labour	61.54	30.77	7.69	76.92	23.08
Marginal farmer cum agricultural labour	72.73	18.18	9.09	72.73	27.27
Marginal farmer	35.71	46.43	17.86	89.29	10.71
Small farmer	18.18	81.82	-	81.82	18.18
Medium farmer	-	100.00	-	100.00	-
Big farmer	-	100.00	-	100.00	-
All	42.42	46.97	10.61	83.33	16.67

Source: Field Survey, 2016-2018.

In case of crop loss, instead of the lessee cultivators, lessors get the compensation from the government. As this tenancy is informal as well as illegal, no document for tenancy is provided to the tenant. So, lessees face the total loss but lessors get rent of land from the lessee as well as compensation from the government, without investing anything.

From the ecological point of view, this tenancy is not sustainable because lessees use more fertilizer on leased-in land to get maximum returns within a short time period. Thus, in the long term, this type of tenancy may cause soil erosion.

Conclusion

Short-term FRT is the most common form of tenancy in present day India. As this type of tenancy is highly risky and to some extent exploitative, steps should be taken so that lessee farmers can avail formal sector loans easily and get compensation for crop loss. In this connection, it may be said that implementation of the model land leasing act, 2016 may create more problems for the lessees. Though this act aims at betterment of lessees as well as lessors, landowners may stop leasing-out fearing adverse impact like *Operation Barga* in future. Already, rumour has spread in Purba Bardhaman and Birbhum districts of West Bengal that the government is going to record the names of lessee cultivators and thus many landowners have stopped leasing-out land for *boro* paddy cultivation. Therefore, steps should be taken to eradicate this type of fear among lessors before legalizing it. Reduction of input cost in paddy cultivation as well as enhancement of crop diversification to make the agricultural system more profitable is highly necessary. All types of farmers should also be trained to make agriculture more sustainable. Moreover, steps should be taken to create job opportunities in the non-farm sector to reduce pressure on land.

9 Living with Crisis

Introduction

In this grounded research, it has been observed that there are many problems in the agriculture sector of the study area, even though the area is known as an agriculturally developed part of Purba Bardhaman district. Major agricultural problems of the study area have been identified and analyzed based on a multi-disciplinary approach. The study reveals that the nature and dimensions of these problems are different. There is an urgent need for addressing these problems. However, no easy solutions to all these problems are available. Therefore, farmers of the study area are in a deeper 'crisis'. As farmers of such a developed area are in a distressed situation, one can assume the situation of the farmers in other parts of the state. In this context, it is important to note that the impacts of the agrarian crisis are not the same on the different categories of farmers.

Though earlier chapters discuss various problems and issues of agriculture and of farmers, there is a need to present the whole matter in a nutshell. The major changes noted in this research are as follows:

Major Shifts in Agrarian Economy

1) Paddy is the most important crop in the study area. However, the areas under *aus* paddy, *boro* paddy, and mustard have decreased considerably, and the area under potato cultivation has increased.
2) There is an adequate supply of seeds, chemical fertilizers, and pesticides in the study area. The most important input-

related problem is the lack of sufficient and timely supply of canal irrigation water. The scope of deep tubewell irrigation is also diminishing day by day because of the declining groundwater table. Due to irrigation-related problems, the region is losing its glory for *boro* paddy cultivation.

3) Farmers use only HYV seed varieties now for paddy cultivation. However, many traditional seed varieties were in use before 1995. Low productivity and double cropping system caused the replacement of traditional seed varieties with the HYV seeds. For paddy, as well as for potato cultivation, farmers still use TL seeds to save the cost of certified seeds. The market is the most important source of seeds. Though government institutions such as seed farms, CRSMF, and KVK also sell seeds to the farmers, but there are many limitations.
4) Chemical fertilizer use is not excessive. In *aman* paddy cultivation, nitrogen use is higher than the prescribed level, but potash use, in most cases, is lower than the prescribed level. During *boro* paddy cultivation, fertilizer use is below the prescribed limit. Most of the farmers have minimal idea about the prescribed limit of N, P, and K use. However, their use of fertilizers is guided by the fertilizer dealers or sellers.
5) Farmers are becoming aware of the negative consequences of chemical fertilizer use on soil fertility. Therefore, the use of micro-nutrients and vitamins is increasing. However, the use of manure and oil-cake has decreased over time with the implementation of Green Revolution technologies.
6) Though the area has a higher percentage of agricultural labours, a labour crisis prevails in most parts of the area during the peak season of paddy plantation and harvesting. There is a lack of male workers for agricultural activity. As agricultural labours do not get jobs throughout the year, most of them migrate for work in non-farm sectors. New generations also do not like to be engaged in agricultural activity, considering it less prestigious and more laborious. Rapidly rising wages

compared to the price growth of agricultural products is also a major problem for the farmers. Though from the economic point of view, decline of agricultural labours is a good sign for development, it is a problem for the farmers as paddy cultivation in the study area is still labour intensive. Recently, the labour crisis and high wages have resulted in increasing harvester use. It has also created a problem for the agricultural labours who still depend on wage income for household expenditure.

7) Though more emphasis is being given on farm mechanization, yet it is not the only solution to the increasing cost of cultivation or the labour crisis. It is revealed that tractorization has created many problems for farmers. Tractor use is indirectly responsible for declining manure use and growing chemical fertilizer use. Harvester use and its resultant effects also have severe environmental impacts in the absence of a proper crop residue management system. Recent efforts for achieving sustainability in agriculture have also been hampered by these machines. Recently, various technologies are being developed aiming at sustainable agriculture. However, use of these technologies is at the initial stage. Farmers are reluctant to adopt these technologies due to various constraints.

8) There are government institutions at different administrative levels, and various programmes for helping farmers. There are schemes for providing financial and non-financial support to the farmers. However, the study reveals that most of the landless lessees, sharecroppers, small, and marginal farmers remained deprived of the benefits of these schemes. Moreover, the extent of government support is inadequate in comparison to the demand. There is also a lack of communication between the farmers and government offices meant for agricultural development, mainly because of the insufficient number of KPS at the ADA offices.

9) Recent government initiatives such as compensation for crop loss and procurement of paddy at MSP are quite good for the farmers, but due to some limitations, farmers

of all categories are not benefitting fully from these schemes.

10) Profitability of crop cultivation has declined over time. A considerable increase in the prices of various cost items and lower rate of increase in the MSP and market price of the agricultural products are the major causes for declining profitability. Cost analysis as followed by the experts proves that paddy cultivation is loss-making, while mustard and potato cultivation are also not profitable enough if A2+FL cost and C2 cost concepts are considered. Moreover, for most years, farmers realize lower prices than the MSP.

11) Though from the economic point of view agriculture is not a profitable activity, farmers are still engaged in cultivation. Such behaviour can be explained by three factors. Firstly, farmers do not consider all cost items in their cost calculations. Secondly, they are not aware of all cost items and the procedures of cost calculations. Therefore, crop cultivation does not seem as loss-making as it is considered by the experts. Besides, there is a lack of suitable and gainful job opportunities in the non-farm sectors. Thirdly, crop cultivation helps farmer households to meet their minimal needs for food and clothing. However, farmers do not like agriculture as their main livelihood option, and most of them do not wish their next generation to be farmers or to be engaged in farming.

12) Various kinds of social, cultural, and behavioural changes have taken place in the last two decades. Most of the farmer households have lost their self-sufficiency in terms of foodstuffs needed. Farmers perceived that modernization and consumerism have resulted in growing dependence on the market, as well as an increase in household-level expenditure. The high rate of inflation is also responsible for growing expenditure. Therefore, to cope with the growing expenditure and fulfil the family's aspirations, most farmer households try to have more than one source of income. Insufficient and declining income from crop cultivation and declining land amount per family is also responsible for forced

migration. In the period of 'jobless growth', members of farmer households are involved in under-paid, petty, and insecure jobs to sustain their family.

13) The increasing short-term tenancy is also a result of the present agrarian crisis. Due to the lack of profitability from crop cultivation and the high wage rate, many farmer households are reluctant to take the risk, and they are leasing-out their land for one or two crop seasons. Contrary to this, farmer households who own tractors and/or submersibles, and farmer households with sufficient family labour are leasing-in land. As this type of short-term tenancy is informal, farmers cannot avail of formal sector loans or compensation for crop loss. Therefore, poor informal lessee cultivators suffer the most.

14) Farmer distress is not a new thing. They are one of the most deprived sections of society from the historical past. At present, they are politically less organized. However, the situation of all farmer households is not the same. Among the various farmer household categories, medium and big farmer households are in a comparatively better position.

Multi-Dimensional Agrarian Crisis

From the literature and field studies, it is clear that the present agrarian crisis is the result of various changing factors. However, it is tough to rightly portray the nature of agrarian crisis using a single perspective. There are multiple dimensions to the prevailing crisis.

The agrarian crisis may be described from the perspective of indebtedness among farmer households and inefficiency in government support mechanisms. Indebtedness among farmer households in the study area is not very high, but a percentage of farmer households still depend on informal sources of loan. Farmers in the study area are not satisfied with the irrigation system and other government supports provided to them.

At present, most of the researchers, experts, scientists,

planners, policy-makers, and farmers perceive that inefficiency in the Green Revolution-led agrarian development model is the most important cause for the creation of multiple dimensions of the agrarian crisis. Proponents of agro-ecology opine that the Green Revolution encouraged selection of such crops which were 'misfit' for many regions from the agro-ecological point of view. Like in other agriculturally developed areas of the country, groundwater table in the study area is declining. The farmers have also perceived that fertilizer and pesticide use has resulted in depletion of soil fertility and increasing pest attacks. Agro-ecologists, as well as supporters of sustainable agriculture, are worried about excessive crop concentration, and use of high yielding seed varieties. However, the adoption of an agro-ecological approach is not a very easy task for the farmers. A complete shift from production-based agriculture to agro-ecology based agriculture may hamper crop production. In the study area, though farmers talked about the negative impacts of chemical pesticides or the effectiveness of bio-fertilizers, no one can think of cultivation without the use of chemical fertilizers or chemical pesticides.

The agrarian crisis may also be described from the viewpoint of cost and returns. From the fourth chapter, it is observed that paddy cultivation in West Bengal is not remunerative enough due to rapidly increasing input costs compared to the realized prices and the MSP. There are three ways to make crop cultivation profitable to the farmers: through a considerable hike in MSP or market price; through input cost reduction; and through productivity growth. However, due to WTO-AoA provisions, an increase in MSP of major crops like paddy or wheat is becoming difficult for the government. Arresting the rising input cost is also not easy for the same reasons. Some of the new technology interventions have potential for input cost reduction. However, these technologies are either out of reach of most small and marginal farmers, or there are also some other problems related to using these technologies. Production increase would not be a solution in the absence of proper demand for farm products and adequate infrastructure (such as storage facilities and a well-developed food processing industry). As farmers of the study area have experienced, market prices remain very low in the years of bumper crop production.

Another dimension of the agrarian crisis is related to economic liberalization and spread of consumerism. There is enough proof that one-sided provisions favouring developed countries have been damaging Indian agriculture directly and indirectly. The Indian government's commitments on subsidy reduction, strict quality control measures and special safeguard mechanisms imposed by the developed countries, and dumping of highly subsidized crops are the significant threats to our agriculture and farmer community. Apparently, liberalization and consumerism are helping in improving the standard of living by supplying various goods and services at comparatively lower prices. However, farmers are becoming more market-dependent and losing their self-reliance due to economic liberalization, and spread of consumerism is resulting in growing expenditure and declining household savings.

Cumulative effects of rising input costs and lower returns, imperfect market competition, unfavourable provisions under WTO-AoA, agro-ecological problems caused by the Green Revolution, lack of proper policy perspective for the farmers' welfare, and growing consumerism have made the farmers' life terrible in the present period of jobless growth. Forced migration, over-dependence on under-paid informal jobs, agriculture for sustenance, and growing informal tenancy are the reflections of the 'crisis' in farmer society.

At present, almost all mainstream political parties have farmers' organizations. However, being poor and less organized, most of these parties are less interested in solving the farmers' problems whole-heartedly; rather, they try to deny or suppress the agrarian problems. Sometimes, they follow a 'cosmetic' development approach. Moreover, as it has already been discussed, there are classes within farmer society, and the powerful section grabs most of the opportunities provided by development programmes, while a majority of subaltern farmers do not get anything.

In this connection, it should be kept in mind that though all farmer categories are facing an agrarian crisis, landless lessee farmers and informal tenant farmers from resource-poor small and marginal categories are more distressed. In most cases, the problems are linked to one another, leading to a higher level of crisis. The study reveals that many initiatives have been taken to address agricultural problems in an isolated manner. New schemes have been launched to promote sustainable agricultural practices. However, most initiatives need to be integrated and properly implemented. Therefore, the time has come to start implementing rigorous multi-pronged strategies for achieving ecological, economic, social, and political sustainability in agriculture with state sponsorship.

Bibliography

(2007). *17th All India Livestock Census*, "Agriculture Implements & Machinery, and Fishery Statistics West Bengal, Bardhaman District", volume 1.

Abrol, I.P., & Sangar, S. (2006). "Sustaining Indian Agriculture-Conservation Agriculture the Way Forward". *Current Science*, 1020-1025.

Adelman, I. & Dalton, G. (1971). "A Factor Analysis of Modernisation in the Village India". *The Economic Journal*, 81(323), p.563-579. Retrieved on 4/12/2019, from https://www.jstor.org/stable/2229849?read-now=1&seq=1#page_scan_tab_contents

Aerthayil, M. (2008). "Agrarian Crisis in India is a Creation of the Polity of Globalisation". *Mainstream*, Vol. XL VI, No. 13.

Aeschbach-Hertig W. and Gleeson T. (2012). "Regional Strategies for the Accelerating Global Problem of Groundwater Depletion", *Nat. Geo. Sci.*, 5, 853-61.

Ahmed, S. (2011). "Determinants of the Choice of Agricultural Tenancy Contracts in Rural Bangladesh". Paper presented at The Australian Agricultural and Resource Economics Society. (AARES) Conference, Melbourne, Australia 8-11 February, Retrieved from https://ageconsearch.umn.edu/bitstream/100529/2/Ahmedpercent20S.pdf.

Ahoy, P. (2016). "Agrarian Crisis India: An Aftermath of the New Economic Reforms". *Asian Journal of Social and Human Sciences*, (2), 31-38.

Alagh, Y.K. (2013). "*The Future of Indian Agriculture*". New Delhi, India: National Book Trust, India.

Ali, S.Z., Sidhu, R.S. & Vatta, K. (2012). "Effectiveness of Minimum Support Price Policy for Paddy in India with a Case Study of Punjab". *Agricultural Economics Research Review*, 25 (347-2016-17003), 231-242.

"*Alu nosto, himghorer bhara niye kshubdho chasi*" [Wastage of Potato, Farmers' Anger due to Charge of Cold Storage]. (2017, March 27). *Anandabazar Patrika*, p.11.

Amarnath, T. & Prasad, A.R. (2009). "Agriculture Development in India since Independence: A Study on Progress, Performance and Determinant". *Journal of Emerging Knowledge on Emerging Markets*, 1.

Arora, V.P.S. (2013). "Agricultural Policies in India: Retrospect and Prospect". *Agricultural Economics Research Review*, 26(2), 135-157.

Athreya, V. (2013). "The Current Agrarian Crisis in India: An Overview". *e-Marxist*, 29(3).

Bakshi, M. (1987). "*Effects of Transition from Feudalism to Capitalism on Rural Bengal 1929 to 1980*" (Ph.D Thesis, University of Calcutta, West Bengal, India). Retrieved from http://shodhganga.inflibnet.ac.in/handle/10603/161765

Bandyopadhyay, D. (2003). "Land Reforms and Agriculture: The West Bengal Experience". *Economic and Political Weekly*, 879-884.

Bandyopadhyay, P. (2013). *Land and All That, Land and Land Reforms in West Bengal: Perspective India*, Kolkata, India: Dey's Publishing.

Banerjee, A., Gertler, P.J. & Ghatak, M. (2002). "Empowerment and Efficiency: Tenancy Reform in West Bengal". *Journal of Political Economy*, 110(2), 239-280. Retrieved from https://www.jstor.org/stable/pdf/10.1086/338744.pdf?refreqid=excelsior%3Aad6b904f60d7f54e4f02bd784 6996743

"Bangla Rolls Out Own Crop Insurance Scheme" (July 23,

2019). *The Hindu Business Line*. Retrieved from https://www.thehindubusinessline.com/economy/agri-business/bengal-rolls-out-own-crop-insurance-scheme/article28690197.ece#.

Banka, A., Singh, A., Malhotra, A., Ramesh, K., Gajjar, J. & Gupta, A. (2021). "Indian Agriculture Since Independence". *International Journal of Research in Engineering, Science and Management*, 4(9), 251-254.

Bansal, S. & Grover, D.K. (2019). "Dynamics of Tenant Farming in India". *Indian Journal of Economics and Development*, 15(2), 225-232.

"Bengal government to pay compensation to farmers." (2017, July17). Retrieved from https://english.kolkata24x7.com/bengal-govt-pay-compensation-farmers.html/.

"Bengal Launches Seed Mission" (2014, June 27). *Business Standard*. Retrieved from https://www.business-standard.com/article/news-ians/bengal-launches-seed-mission-114062700669_1.html

"*Bhagchasir opomrityu Purbosthalite*" [Unnatural death of sharecropper in Purbasthali]. (2018, December 6). *Anandabazar Patrika*, p.A2.

Bhalla, G.S. & Singh, G. (2009). "Economic Liberalisation and Indian Agriculture: A State Wise Analysis". *Economic and Political Weekly*, 44(52), 34-44. Retrieved from http://www.environmentportal.in/files/economic.pdf.

Bhalla, G.S. (2004). "*Globalisation and Indian Agriculture*", New Delhi, India: Academic Foundation.

Bhattacharyya, S. (2016). "'Globalized' and' Decoupled': Indian Economy in the Era of Global Crisis". *Social Scientist*, 44(7/8), 53-77.

Bhaumik, S.K. (1993). *Tenancy Relations and Agrarian Development: A Study of West Bengal*. New Delhi, India: Sage.

Bhoi, B.K. & Dadhich, C.L. (2019). "Agrarian Distress in India: Possible Solutions". *Indira Gandhi Institute of Development Research, Mumbai, India*, WP, 17, 1-13.

Bhue, C. & Kikon, R. (2020). "Issues and Challenges of MSP as an Income Enhancement Approach in India". *The Indian Economic Journal*, 68(2), 290-295.

Bhukta, A. (2007). *Biswayan o Bharater Krishi Arthaniti*. Kolkata, India: Grantha Mitra.

Bihari, B.A.N.K.E.Y., Singh, M.A.D.A.N., Bishnoi, R. & Mishra, P.K. (2019). "Issues, Challenges and Strategies for Doubling the Farmers' Income in India-A Review". *Indian Journal of Agricultural Sciences*, 89(8), 1219-1224.

Binswanger, H.P. (1978). "The Economics of Tractors in South Asia: An Analytical Review". *New York/India: Agricultural Development Council, New York and International Crop Research Institute for the Semi-arid Tropics, Hyderabad.*

Biru, B.K. & Barpujari, I. (n.d.). "Jan Sunwai on The Present Agrarian Crisis". *Gene Campaign*, 1-42. Retrieved on February 8, 2019, from https://genecampaign.org/wp content/uploads/2014/07/THE_AGRARIAN_CRISIS_IN_INDIA.pdf

Boyce, J.K. (1984). "Agricultural Growth in West Bengal, 1949-50 to 1980-81: A Review of the Evidence", *Economic and Political Weekly*, 19(13), A9-A16.

Brown, T. (2018). *Farmers, Subalterns and Activists: Social Politics of Sustainable Agriculture in India*, Cambridge, UK: Cambridge University Press.

"Budget 2019: Modi Govt's New Scheme Offers Rs. 6000 Directly into Farmers' Bank Accounts." (2019, February 1). *Business Today*. Retrieved on February 8, 2019 from https://www.businesstoday.in/budget-2019/news

Bureau of Applied Economics and Statistics. (Various years). *District Statistical Handbook*, Government of West Bengal.

Census of India (2011). District Census Handbook, Barddhaman.

Centre for Study of Developing Societies. (2018). State of Indian Farmers: A Report. 1-36. Retrieved from http://www.lokniti.org/pdf/Farmers_Report_Final.pdf

Chakrabarti, A. (2013). "Trends in Agricultural Productivity in Post Land Reform Period a Study of the Impact of Agricultural Productivity on Employment and the Economy of West Bengal" (Ph.D Thesis, University of North Bengal, West Bengal, India). Retrieved from http://shodhganga.inflibnet.ac.in/handle/10603/165847

Chakrabarti, B. (1995). "Economic Development of the District of Burdwan and Bankura 1951-81: A Comparative Study". (Ph. D. Thesis, Department of Economics, The University of Burdwan, West Bengal, India). Retrieved from http://hdl.handle.net/10603/65820.

Chakraborty, S. & Sarkar, P. (2018, December, 13). "Twin Thrones Sprout for Bengal Paddy Farmers". *The Telegraph*, p.8.

Chand, R. & Haque, T. (1998). "Rice-Wheat Crop System in Indo-Gangetic Region: Issues Concerning Sustainability". *Economic and Political Weekly*, 33(26), A108-A112.

Chand, R. & Pandey, L.M. (2008). "Fertilizer Growth, Imbalances and Subsidies: Trends and Implications", NPP Discussion Paper 02/2008, *National Centre for Agricultural Economics and Policy Research: New Delhi*, 1-41. Retrieved from http://www.ncap.res.in/upload_files/others/oth_13.pdf

Chand, R. & Parappurathu, S. (2012). "Temporal and Spatial Variations in Agricultural Growth and It's Determinants". *Economic and Political Weekly*, 46(26 & 27), 55-64.

Chand, R., Lakshmiprasanna, P.A. & Singh, A. (2011). "Farm Size and Productivity: Understanding the Strength of Smallholders and Improving their Livelihoods". *Economic and Political Weekly*, 46 (26 & 27), 5-11.

Chand, R., Saxena, R. and Rana, S. (2015). "Estimates and Analysis of Farm Income in India: 1983-84 to 2011-12", *Economic and Political Weekly*, Vol. 40, No. 22, May 28, pp. 139-145.

Chand, R., Saxena, R. and Rana, S. (2015). "Estimates and Analysis of Farm Income in India: 1983-84 to 2011-12", *Economic and Political Weekly*, 50(22), 139-145.

Chand, R., Saxena, R. and Rana, S. (2015). "Estimates and Analysis of Farm Income in India: 1983-84 to 2011-12", *Economic and Political Weekly*, 139-145.

Chandran, S. R (2017). "Impact of Mechanization on Cost of Cultivation and Production of Paddy in Thrissur". *International Journal of Applied Research*, 3(10), 79-81.

"*Chaser jol niye montrir samne kshov chasider*" [Farmers' Anger in Front of Minister due to Lack of Irrigation Water]. (2015). *Anandabazar Patrika*, p.17.

"*Chasider bikhobh*" [Farmers' Agitation]. (2017, February 11). *Anandabazar Patrika*, p.10.

"*Chasir aswavabik mrityu, chase kshotir aviyog*" [Unnatural Death of Farmer, Claim for Crop Loss]. (2014). *Anandabazar Patrika*, p.4.

Chattopadhyay, A.K. (2005). "Distributive Impact of Agricultural Growth in Rural West Bengal". *Economic and Political Weekly*, 40(53), 5601-5610. Retrieved from https://www.jstor.org/stable/pdf/4417614.pdf?refreqid=excelsior%3A2b72065996177470ac81b6af6b92b85e

Chaudhary, R. (2018). "Crop Residue Burning: A Case of On-Farm Usage of Crop Residue". *Economic and Political Weekly*, 53(46), 60-61.

Chauhan, N. (2021). 'Minimum Support Price: All that Glitters is not Gold". https://sprf.in/wp-content/uploads/2021/11/SPRF-2021_Comm_Minimum-Support-Price-1.pdf

Crabb, P.B. & Stern, S.E. (2010). "Technology Traps: Who is Responsible?" *International Journal of Technologies*, 1(2), 19-26. Retrieved from https://www.researchgate.net/publication/220256918_Technology_Traps_Who_Is_Responsible

CSDS. (2018). "State of Indian Farmers: A Report". 1-36. Retrieved from http://www.lokniti.org/pdf/Farmers_Report_Final.pdf

Dalin, C., Wada, Y., Kastner, T. & Puma, M.J., (2017). "Groundwater Depletion Embedded in International Food Trade". Nature 543, 700-704. https://doi.org/10.1038/nature21403.

Dangar, S., Asoka, A. & Mishra, V. (2021). "Causes and Implications of Groundwater Depletion in India: A Review". *Journal of Hydrology*, 596, 126103.

Das, R. (2020). "Minimum Support Price in India: What Determines Farmers' Access?". *Agricultural Economics Research Review*, 33(347-2020-1416).

Das, T. (2012). "Economic Liberalisation and Agrarian Crisis in India". *International Journal of Applied Research in Business Administration and Economics*, 1(1).

Dasgupta, S. & Bhaumik, S.K. (2014). "Crop Diversification and Agricultural Growth in West Bengal", *Indian Journal of Agricultural Economics*, 69(1), 107-124.

Deokar, B. & Shetty, S.L. (2014). "Growth in Indian Agriculture: Responding to Policy Initiatives since 2004-05", *Economic and Political Weekly*, 49 (26&27), 101-104.

Department of Planning, Statistics & Programme Monitoring. Economic Review 2016-17, 2017-18.

Deshpande, R.S. (2002). "Suicide by Farmers in Karnataka: Agrarian Distress and Possible Alleviatory Steps". *Economic and Political Weekly*, 2601-2610.

Deshpande, R.S., Bhende, M.J., Thippiah, P. and Vivekananda, M. (2004). "State of The Indian Farmer: Crops and Cultivation", *Academic Foundation*, New Delhi.

Deshpande, T. (2017). "State of Agriculture in India". *PRS Legislative Research*. Retrieved on October 23, 2019, from https://www.prsindia.org/policy/discussion-papers/state-agriculture-india

Dev, S.M. & Rao, N.C. (2010). "Agricultural Price Policy, Farm Profitability and Food Security". *Economic and Political Weekly*. 45(26-27), 174-182.

Dey, C.K. & Mistri, T. (2018). "Changing Trends of Crop Concentration in Purba Bardhaman District, West Bengal-A Comparative Analysis". *International Journal of Applied Social Science*, 5(6), 790-799. Retrieved from http://scientificresearchjournal.com/wp-content/plugins/download-attachments/includes/download.php?id=4263

"*Dhan bikrir bochor par, taka na melai kshov*" [One year over of Paddy Selling, Fury due to No Payment]. (2017, January 18). *Anandabazar Patrika*, p.11.

"*Dhane kshoti, fer chasir mrityu Bhatare*" [Loss in Paddy, Death of Farmer in Bhatar Again]. (2017, May 5). *Anandabazar Patrika.*

Dhas, A.C. (2009). "Agricultural Crisis in India: The Root Cause and Consequences". *Munich Personal RePEc Archive*, paper no. 18930, 1-14. Retrieved from https://ideas.repec .org

Directorate of Animal Resources and Animal Health. (2003). *17th All India Livestock Census, Agriculture Implements & Machinery, and Fishery Statistics West Bengal, Bardhaman District*, Vol. 1.

Directorate of Economics and Statistics. (n.d.). "Cost of Cultivation/Production and Related Data". Retrieved from https://eands.dacnet.nic.in/Cost_of_Cultivation .html

Durga, A.R. & Kumar, D.S. (2013). "Economic Analysis of the System of Rice Intensification: Evidence from Southern India". *The Bangladesh Development Studies*, 36(1), 79-93. Retrieved from https://bids.org.bd/uploads/publication/BDS/36/36-1/03_Economic%20Analysis%20 of%20 System.pdf

Elaoud, A. & Chehaibi, S. (2011). "Soil Compaction due to Tractor Traffic". *Journal of Failure Analysis and Prevention*, 11(5), 539-545. Retrieved from https://www.researchgate.net/publication/225541728_Soil_Compaction_Due_to_Tractor_Traffic

Ellenberg, G.B. (2000). "Debating Farm Power: Draft Animals, Tractors and The United States Department of Agriculture". *Agricultural History*, 74(2), 545-568. Retrieved from https://www.jstor.org/stable/3744871?seq=1#metadata_info_tab_contents

Eric, O., Prince, A.A. & Elfrada, A.N.A. (2014). "Effects of Education on the Agricultural Productivity of Farmers in the Offinso Municipality". *International Journal of*

Development Research, 4(9), 1951-1960. Retrieved from https://www.researchgate.net

FAI. (2020). Fertiliser Statistics 2019-20, 65th Edition. The Fertiliser Association of India, New Delhi.

"*Fer chasir apomrityu*" [Unnatural death of farmer again]. (2017, 18 May). *Anandabazar Patrika*, p.12.

"*Fer dhan chasir mrityu Montesware*" [Again paddy farmer's death in Monteswar]. (2015, August 27). *Anandabazar Patrika*, p.13.

Fofack, H. (2008). "Technology Trap and Poverty Trap in Sub-Saharan Africa". Policy Research Working Paper 4582. Retrieved on 19/12/19, from http://documents.worldbank.org/curated/en/169021468198874707/pdf/wps4582.pdf.

Food and Agriculture Organization (n.d.). Retrieved from http://www.fao.org/india/fao-in-india/india-at-a-glance/en/

Frey, C.B. & Osborne, M.A. (2013). "The Future of Employment: How Susceptible are Job to Computerisation?" Retrieved on 19/12/19, from https://www.oxfordmartin.ox.ac.uk/downloads/academic/The_Future_of_Employment.pdf

Gaurav, S. & Mishra, S. (2011). "Size Class and Return to Cultivation: A Cold Case Reopened". *Indira Gandhi Institute of Development Research*, Mumbai. 1-24. Retrieved on August 27, 2018, from http://www.igidr.ac.in/pdf/publication/WP-2011-027.pdf

Ghosal, S. (2018). "West Bengal: Paddy Procurement Scheme Witness 8 Time Increase". *The Economic Times*. Retrieved from https://economictimes.indiatimes.com

Ghosh, B.K. (2008). "Growth and Cropping Pattern Changes in West Bengal during 1970-71 to 2005-06", (Ph.D. Thesis, The University of Burdwan, West Bengal, India). Retrieved from http://hdl.handle.net/10603/61970

Ghosh, B.K. (2010). "Growth and Variability in the Production of Crops in West Bengal Agriculture". *Trends in Agricultural Economics*, 3(3), 135-146. Retrieved from http://docsdrive.com/pdfs/ansinet/tae/2010/135-146.pdf

Gill, S.S. (2018). "Crop Residue Burning: Solution Marred by Policy Confusion". *Economic and Political Weekly*, 53(36), 23-25.

Gneiting, U. & Sonenshine, J. (2018). "A Living Income for Small-Scale Farmers: Tackling Unequal Risks and Market Power", *Oxfam Discussion Paper*. Oxford: Oxfam International. 1-24. Retrieved from https://oxfamilibrary.openrepository.com/bitstream/handle/10546/620596/dp-living-income-smallscale-farmers-151118-en.pdf

Gopal, S. (1949). The Permanent Settlement in Bengal and its Results, *London: Allen & Unwin*, Retrieved from http://203.200.22.249:8080/bitstream/123456789/13537/1/G5072.pdf

Goswami, B. & Bezbaruah, M.P. (2013). "Incidence, Forms and Determinants of Tenancy in the Agrarian Setup of the Assam Plains", *Economic and Political Weekly*, 47(42), 60-68.

Gould, C.F. & Urpelainene, J. (2018). "LPG as a Clean Cooking Fuel: Adoption, Use and Impact in Rural India". *Energy Policy*, 122(2018), 395-408. Retrieved from https://static1.squarespace.com/static/53856e1ee4b00c6f1fc1f602/t/5b84722303ce648b686e5c25/1535406630658/Gould+Urpelainen+2018+LPG.pdf

Gulati, A., Terway, P. & Hussain, S. (2018). "Crop Insurance in India: Key Issues and Way Forward". *Indian Council for Research on International Economic Relations*. Working paper 352. p.1-55. Retrieved from https://www.google.com/search?q=Agricultural+insurance+in+india.pdf&rlz=1C1GIGM_enIN838IN838&oq=Agricultural+insurance+in+india.pdf&aqs=chrome..69i57j0l3.16185j0j8&sourceid=chrome&ie=UTF-8#.

Gururaj, B., Hansa, K.R., Ramesh, K.R. & Mahadevaiah, G.S. (2017). "Doubling of Small and Marginal Farmers Income through Non-Farm and Farm Sector in Karnataka". *Economic Affairs*, 64(4), 581-587. Retrieved from https://ndpublisher.in/admin/issues/EAv62n4b.pdf

Haldar, S., Honnaiah & Govindaraj, G. (2012). "System of Rice Intensification (SRI) Method of Rice Cultivation in West Bengal (India): An Economic Analysis". *International Association of Agricultural Economists (IAAE) Triennial Conference*, Foz do Iguaçu, Brazil, 18-24 August, 2012. Retrieved from http://ageconsearch.umn.edu/bitstream/126234/2/IAAE%20conference%20paper%20on%20SRI-uploading.pdf

Hanstad, T. (2005). *Improving Land Access for India's Rural Poor-A Review of Past Efforts and Recommendations for the Road Ahead, Rural Development Institute*, 1-44.

Haque, T (1996). *Sustainability of Small Holder Agriculture in India, Concept Publishing Company*, New Delhi.

Haque, T. (2012). Impact of land Leasing Restrictions on Agricultural Efficiency and Equity in India, New Delhi, Concept Publishing Company.

Haque, T. (2021). "Reforming and Transforming Indian Agriculture: Key Challenges and Pathways". *Agricultural Economics Research Review*, 34 (347-2022-539).

Harttgen, K., Klasen, S. & Vollmer, S. (2011). "An African Growth Miracle Or: What Do Asset Indices Tell Us about Trends in Economic Performance?" 1-32. Retrieved from http://iariw.org/papers/2011/harttgenpaper.pdf

"India's Consumption Story: Why You Just Can't Ignore This Opportunity". (2018, October 24). *Economic Times*. Retrieved on October 24, 2019, from https://economic times.indiatimes.com

Indian Council on Food and Agriculture. (2016). ICFA report on Crop Insurance in India, 2016. Retrieved from https://www.google.com/search?q=agricultural+insurance+in+west+bengal+history&rlz=1C1GIGM_ enIN838IN838&ei=OmJrXfLYKcTjvgTpgrfYCg&start=10 &sa=N&ved=0ahUKEwjy3e_c_K7kAhXEsY8KHWnBDasQ8NMDCL8B&biw =1366&bih=657#

Jagannathan, R. (2018). "The Jobs Crisis in India". New Delhi, India: Macmillan.

Jain, M., Fishman, R., Mondal, P., Galford, G.L., Bhattarai, N., Naeem, S., Lall, U. and Defries, R.S. (2021). "Groundwater Depletion will Reduce Cropping Intensity in India", *Sci. Adv.*, 7, eabd2849.

"*Jamalpure chasir mrityu, nalish kshotir*" [Farmer's Death in Jamalpur, Complain of Crop Loss]. (2017). *Anandabazar Patrika*, p.10.

Jebaraj, P. (2018, December 8). "Why The Farmers All Over India on The Streets?", *The Hindu*, Retrieved from https://www.thehindu.com/news/national/why-are-farmers-all-over-india-on-the-streets/article25698970.ece

Jha, P. (2007). "Some Aspects of the Well-being of India's Agricultural Labour in the Context of Contemporary Agrarian Crisis". *Indian Journal of Labour Economics*, 49(4), 741-64.

Jodhka, S. (2016). "Revisiting the Rural in 21st Century India". *The Economic and Political Weekly*, 51 (26&27), p.5-7.

Jodhka, S.S. (2012). "Agrarian Changes in The Times of (Neo -Liberal) "Crises": Revising Attached Labour in Haryana". *Economic and Political Weekly*, 47 (26&27), 5-13.

Joshi, P.C. (1974). "Land Reform and Agrarian Change in India and Pakistan since 1947: II", *The Journal of Peasant Studies*, 1:3, 326-362. https://doi.org/10.1080/03066157408437894

Kalamkar, S.S. & Shroff, S. (2011). Impact of Rehabilitation Package in Suicide Prone Districts of Vidarbha. *Economic and Political Weekly*, 46(05), 10-16.

Kar, N. & Banerjee, A. (2010). *Changing Scenarios of Agricultural System and Its Impact on Rural Society in West Bengal during the Left Front Regime: A Study of Two Villages (1977-2006)* (Ph.D. thesis, Netaji Subhas Open University, West Bengal, India). Retrieved from http://shodhganga.inflibnet.ac.in/handle/10603/2970

Kaur, H. & Dhaliwal, N.K. (2018). Progress of Kisan Credit Card Scheme in India. *Amity Journal of Agri-business*, 3(1), 26-36.

Kennedy, J. & King, L. (2014). The Political Economy of Farmers' Suicides in India: Indebted Cash-Crop Farmers with Marginal Landholdings Explain State-Level Variation in Suicide Rates. *Globalization and Health*, 10(1), 1-9.

Krebstein, K., Janowsky, K. & Reintam, E. (2014). The Effect of Tractor Wheeling on the Soil Properties and Root Growth of Smooth Brome. *Plant Soil Environ*, 60(2), 74-79. Retrieved from https://www.agriculturejournals.cz/publicFiles/114060.pdf

"Krishak Bandhu: Mamata announces Sops for Farmers". (2018, December 31). *The Hindu*. Retrieved on February 8, 2019, from https://www.thehindubusinessline.com

Krishna, V.V., Aravalath, L.M. & Vikraman, S. (2019). Does Caste Determine Farmer Access to Quality Information? *PLoS ONE*, 14(1), 1-22. Retrieved fromhttps://www.researchgate.net/publication/330641842

Krishnaraj, M. & Kanchi, A. (2008). *Women Farmers of India*. New Delhi, India: National Book Trust.

Kumar, A., Kumar, P. & Sharma, A.N. (2011). Rural Poverty and Agricultural Growth in India: Implications for the Twelfth Five Year Plan. *Indian Journal of Agricultural Economics*, 66(3), 269-278. Retrieved from https://pdfs.semanticscholar.org/01ed/ae56bdacfc48fcf3c428737ac3bcaf85137e.pdf

Kumar, A., Sonkar, V.K. & Aditya, K.S. (2022). "Assessing the Impact of Lending Through Kisan Credit Cards in Rural India: Evidence from Eastern India". *The European Journal of Development Research*, 1-21.

Kumar, D.A.D. & Reddy, D.N. (2017). "High Pesticide Use in India: Health Implications". *Health Action*, August, 7-12.

Kumar, S. & Singh, S.K. (2016). "Effect of Hybridisation on Secondary Characteristics of Rice Plant". *Jharkhand Journal of Development and Management Studies*, 14 (2), 7009-7021. Retrieved from https://www.xiss.ac.in/JJDMS/Vol14/Issue2/pdf/5-Kumar%20&%20Singh.pdf

Kumar, S. (2020). "Agrarian Crisis and Framers Issues in India. From Indebtedness to Suicide". *Agrarian Crisis and Farmers Suicide in India.* The Voice Unheard, Best Publishing House: New Delhi, India, 125-154.

Langa, M. & Sriram, J. (2017, June 10). "Farmers' Protest: Fault Lines in the Fields". *The Hindu.* Retrieved on December 4, 2018, from https://www.thehindu.com/news/national/other-states/fault-lines-in-the-fields/article18951520.ece

Lerche, J. (2011). "Agrarian Crisis and Agrarian Questions in India". *Journal of Agrarian Change*, 11(1), 104-118.

Mandal, S. (2015). "Performance, Current Status and Visioning Agriculture in West Bengal".

Mani, G. (2016). "Model Agricultural Land Leasing Act, 2016: Some Observations", *Economic and Political Weekly*, 51(42), Retrieved from https://www.epw.in/journal/2016/42/web-exclusives/model-agricultural-land-leasing-act-2016-some-observations.html

Mani, G. (2016). "Study on Implementation of Kisan Credit Card Scheme". *Department of Economic Analysis and Research, NABARD*, Mumbai. Retrieved from https://www.google.com/search?rlz=1C1GIGM_enIN838IN838&ei=12NvXdmAAcO89QOsm7qwDA&q=kisan+credit+card+scheme.pdf&oq=kisan+credit+card+scheme.pdf&gs_l=psy-ab.3..0i22i30l3.20654.21879..22587...0.2..0.353.1077.2-3j1......0....1..gws-wiz.......0i71.kmafU4gzQ1s&ved=0ahUKEwiZyL7pzrbkAhVDXn0KHayNDsYQ4dUDCAs&uact=5#

McLane, J.R. (1993). *Land and Local Kingship in Eighteenth-Century Bengal.* Cambridge University Press, Cambridge UK.

Mehta, D., Trivedi, H. & Mehta, N.K. (2016). "Indian Kisan Credit Card Scheme: An Analytical Study". *BRAND. Broad Research in Accounting, Negotiation, and Distribution*, 7(1), 19-23.

"*Milche na secher jol, fer bipake chasira*" [Crisis of Irrigation

Water, Farmers are in Trouble]. (2015). *Anandabazar Patrika*, p.7.

Mishra, P. (2013). "Agriculture vs Non-Agriculture". *Economic and Political Weekly*, 17(15), 82-98.

Mishra, S. and Reddy, N. (2008). "Crisis in Agriculture and Rural Distress in Post-Reform India", *India Development Report.*

Mishra, S. (2014). "Farmers' Suicides in India, 1995-2012: Measurement and Interpretation". *London: Asia Research Centre.*

Mishra, S. & Reddy, D.N. (2011). "Persistence of Crisis in Indian Agriculture". *India Development Report 2011*, 48.

Mithra, J. & Bhaskaran, R. (2018). "Comparative Study of System of Rice Intensification and Traditional Method of Rice Cultivation in Thiruvarur District of Tamil Nadu-India". *International.*

Mkzihe, H.P.B. (2013). "The Stunted Structural Transformation of the Indian Economy". *Economic & Political Weekly*, 48 (26&27), 5-13.

Mohanakumar, S. (2014). "*Tenancy Relation in India: Observations From a Field Study*", Working Paper 173, Institute of Development Studies, Jaipur, Rajasthan, India.

Mohanty, B.B. (1999). "Agricultural Modernisation and Social Inequality-Case Study of Satara District". *Economic and Political Weekly*, June 26, 1999, A50-A61.

Mohanty, B.B. (2001). "Suicides of Farmers in Maharashtra: A Socio-Economic Analysis". *Review of Development and Change*, 6(2), 146-189.

Mohanty, B.B. (2005). "'We are Like The Living Dead': Farmer's Suicides in Maharashtra, Western India". *Journal of Peasant Studies*, 32(2), 243-276.

Mohanty, B.K. (2018, December 3). "Why Our Farmers are So Angry". *The Telegraph*, p.1.

Moudgal, S. (2016, December 12). "Mounting Debt, Successive

Drought Push up Number of Farmer Suicide Cases", *The Times of India*, Retrieved from https://timesofindia.indiatimes.com/city/bengaluru/Mounting-debts-successive-droughts-push-up-number-of-farmer-suicide-cases/articleshow/55929928.cms

Mukherjee, S. & Chattopadhyay, S. (2017). "Crop Diversification in West Bengal: A District Level Analysis for the period 1980-81 to 2011-12", *Journal of Rural Development*, 36(4) 501- 530.

Mukhopadhyay, S. Kumar. (1992). "Operation Barga: A Study in Changing Rural Power Structure in West Bengal" (Ph.D. Thesis, University of Calcutta, West Bengal, India). Retrieved from http://shodhganga.inflibnet.ac.in/handle/10603/164715

Murthy, R.V. (2013). "Political Economy of Agrarian Crisis and Subsistence under Neoliberalism in India".

Nadeem, A. (2019, April 10). 'The Jobs Data Mystery: Understanding India's Big Crisis in a Fraught Poll Season'. *The Economic Times*. Retrieved on October 24, 2019, from https://economictimes.indiatimes.com

Nadkarni, M.V. (1988). "Crisis of Increasing Cost in Agriculture: Is There a Way Out?", *Economic and Political Weekly*, 23(39), A114-A119. Retrieved from https://www.jstor.org

Nadkarni, M.V. (1996). "Accelerating Commercialisation of Agriculture: Dynamic Agriculture and Stagnating Peasants?", *Economic and Political Weekly*, June 29, 1996, A63-A73.

Nadkarni, M.V. (2018). "Can It Be Overcome? Crisis in Indian Agriculture". *Economic and Political Weekly*, 53(17).

Nagaraj, K., Sainath, P., Rukmani, R. & Gopinath, R. (2014). "Farmers' Suicide in India: Magnitude, Trends and Spatial Patterns, 1997-2012". *Review of Agrarian Studies*. Retrieved from http://ras.org.in/farmers_suicides_in_india

Nair, S.R. & Eapen, L.M. (2015). "Agrarian Performance and Food Price Inflation in India Pre and Post-Economic

Liberalization", *Economic and Political Weekly*, 50(31), 49-60.

Nandy, D. & Sudhasil, S. (2000). "Trends in West Bengal Agriculture", Encounter, 3(6), 101-119. Retrieved from https://www.researchgate.net/publication/260052588

Nandy, P. (2002). "Agro-based Industries of Burdwan City and Adjoining Areas: A Study with Special Reference to Rice Milling". 1-312. (Ph.D. Thesis, The University of Burdwan, West Bengal, India). Retrieved from http://hdl.handle.net/10603/66051

Narayanamoorthy, A. (2013). "Profitability in Crops Cultivation in India: Some Evidence from Cost of Cultivation Survey Data". *Indian Journal of Agricultural Economics*, 68 (1), 104-121.

Narayanamoorthy, A. (2017). "Farm Income in India: Myths and Realities". *Indian Journal of Agricultural Economics*, 72(1), 49-75.

Narayanamoorthy, A., Ali, P. & Suresh, R. (2014). "How Profitable is Cultivation of Rainfed Crops? Some Insights From Cost of Cultivation Studies". *Agricultural Economics Research Review*, 27 (2), 231-241.

Nath, S.K. (2008). "Manual on Cost of Cultivation Surveys", *Ministry of Statistics and Programme Implementation, Government of India*. Retrieved fromhttp://mospi.nic.in/sites/default/files/publication_reports/manual_cost_cultivation_surveys_23july08_0.pdf

Nayak, P. & Solanki, H. (2021). "Pesticides and Indian Agriculture-A Review". *Int. J. Res. Granthaalayah*, 9, 250-63.

Neog, R. (2018). "Assessing the Impact of Chemical Fertilizers on Soil Acidification: A Study on Jorhat District of Assam, India". *Agric. Sci. Dig.*, 38, 270-274.

Nirmala, A.K. (2003). "Market Imperfections and Farmers' Distress in Andhr Pradesh". *Visakhapatnam: Andhra University.*

NITI Aayog. (2016). *Report of the Expert Committee on Land Leasing*, New Delhi: Government of India.

Patra, D.C. (2012). "The Case of LPG". *Economic & Political Weekly*, 47(49), 77-78. Retrieved from http://www.india environmentportal.org.in

Paul, S. (2016). 1930-1943: Agrarian Transformation and the Famine in Bengal, 1-44, (Ph.D. Thesis, Institute of Economic Research, Hitotsubashi University, Kunitachi,Tokyo, Japan). Retrieved from https://hermes-ir.lib.hit-u.ac.jp/rs/bitstream/10086/28239/1/wp2016-11.pdf

Petterson, J.C.K. (1910). *Bengal District Gazetteers, Burdwan*. Calcutta, India: Bengal Secretrait Book Depot.

Pillai, S.R. (2007). "Agrarian Crisis and The Way Out". *The Marxist*, 23(3), 1-18.

Pimbert, M.P. (2018). "Global Status of Agroecology: A Perspective on Current Practices, Potential and Challenges". *Economic and Political Weekly*, 53(41), 52-57.

Pingle, G. (2011). "Irrigation in Telengana: The Rise and Fall of Tanks". *Economic and Political Weekly*, 46 (26&27), 123-130.

Posani, B. (2009). "Crisis in the Countryside: Farmer Suicides and The Political Economy of Agrarian Distress in India" (No. 09-95). London. Retrieved on February 25, 2018, from http://www.lse.ac.uk/internationalDevelopment/pdf/WP/WP95.pdf

Prasad, C.S. (2006)."System of Rice Intensification in India: Innovation History and Institutional Challenges". *WWF, India and Xavier Institute of Management, Bhubaneshwar*. Retrieved from http://services.iriskf.org/data/articles/Document16122006420.4017603.pdf

Press Information Bureau. (2019). Retrieved from https://pib.gov.in/Pressreleaseshare.aspx?PRID=1573283

Press Trust of India. (2019, July 23).WB Stops Implementing PMFBY; rolls out its own scheme, *Deccan Herald*. Retrieved from https://www.deccanherald.com/amp/

national/wb-stops-implementing-pmfby-rolls-out-its-own-scheme-749073.html

Pushpa., Srivastava, S.K. & Agarwal, P.K. (2017). "Comparative Study on Cost of Cultivation and Economic Returns from Major Crops in Eastern Region of Uttar Pradesh". *International Journal of Agriculture, Environment and Biotechnology*, 10(3), 387-399. Retrieved on September 1, 2018, from http://ndpublisher.in/admin/issues/IJAEBv 10n3p.pdf

Radhakrishna, R. (2002). 'Agricultural Growth, Employment and Poverty: A Policy Perspective". *Economic and Political Weekly*, 243-249.

Radhakrishna, R. (2002). "Agricultural Growth, Employment and Poverty: A Policy Perspective". *Economic and Political Weekly*, 243-250.

Raghavan, M. (2008). "Changing Pattern of Input Use and Cost of Cultivation". *Economic and Political Weekly*, 43 (26-27), 123-129.

Rajendran, Dr. S. (2003). "Environment And Health Aspects of Pesticides Use in Indian Agriculture" in Martin J. Bunch, V. Madha Suresh and T. Vasantha Kumaran, eds., *Proceedings of the Third International Conference on Environment and Health, Chennai, India*, 15-17 December. Chennai: Department of Geography, University of Madras and Faculty of Environmental Studies, York University. Pages 353 - 373.

Raju, S. (1990). "Fertiliser Use and Development of Institutional Infrastructure: A Preliminary Analysis", *Economic and Political Weekly*, June 30, 1990, A79-A85.

Ramakumar, R. (2017). Lessons from Agricultural Debt Waiver and Debt Relief Scheme of 2008. Retrieved from https://rbidocs.rbi.org.in/rdocs/content/pdfs/Dr.%20R.%20Ramakumar_Paper(31082017).pdf

Ranganathan,T. (2015). "Farmers Income in India: Evidence from Secondary Data", *Agricultural Situation in India*, 72 (3), 30-70, Institute of Economic Growth, New Delhi, India.

Rao, C.H.H. (2004). "Preface to Millennium Study". In G.S Bhalla, *Globalisation and Indian Agriculture*, New Delhi, India: Academic Foundation.

Ravikanth, D. (2014). "WTO Upside Down: Trade Facilitation vs Agriculture". *Economic and Political Weekly*, 21-25.

Ravikanth, D. (2014). "WTO Upside Down: Trade facilitation vs agriculture". *Economic and Political Weekly*, 49(36), 21-25.

Ray, S. (2007). "Economic Policy and Agriculture". In Shovan Ray (Ed.), *Handbook of Agriculture in India*, New Delhi, India: Oxford University Press, 34-56.

Ray, S. & Ghosh, D. (2007). "Modern Agriculture and The Ecologically Handicapped: Fading Glory of Boro Paddy Cultivation in West Bengal", *Economic and Political Weekly*, 42(26), 2534- 2542.

Reddy, A. (2021). "Policy Implications of Minimum Support Price for Agriculture in India". *Academia Letters,* Article, 2406.

Reddy, D.N. & Mishra, S. (2008). "Crisis in Agriculture and Rural Distress in Post-Reform India". *India Development Report*, 2008, 40-53.

Reddy, D.N. & Mishra, S. (2010). "Economic Reforms, Small Farmer Economy and Agrarian Crisis". Agrarian Crisis and Farmer Suicides, 43-69.

Reddy, D.N., & Mishra, S. (Eds.) (2010). *Agrarian Crisis in India*. Oxford University Press.

Reddy, N.M., Radhika, K. & Sunandini, G.P. (2009). "Enhancing Rural Farm Resources through Seed Production". *Kurukshetra*, 57 (7).

Rodell, M., Velicogna, I., Famiglietti, J.S., (2009). "Satellite-Based Estimates of Groundwater Depletion in India". *Nature* 460, 999-1002. https://doi.org/10.1038/nature08238.

Rogaly, B. (1999). "Dangerous Liaisons? Seasonal Migration and Agrarian Change in West Bengal," In Ben Rogaly, Barbara Harriss-White and Sugata Bose (eds.) Sonar Bangla? *Agricultural Growth and Agrarian Change in*

West Bengal and Bangladesh, (pp. 357-381). New Delhi, India: Sage Publications.

Rohit, T.K. & M.S. Preetha. (2019, June 16). "When Sugar Leaves a Bitter Taste". *The Hindu*. Retrieved from https://www.thehindu.com/news/national/tamil-nadu/when-sugar-leaves-a-bitter-taste/article27952331.ece

Rola A.C. and Pingali, P.L. (1993). "Pesticides, Rice Productivity and Farmers' Health-An Economic Assessment", *World Resources Institute and IRRI*, Los Banos, Laguna, Philippines.

Roy, B.C., Mondal, B., Ojha, S., Biswas, R.K. & Dutta, V. (2018). "Performance Evaluation of Pradhan Mantri Fasal Bima Yojana (PMFBY) in West Bengal". *Agro-Economic Research Centre, Visva Bharati*. Study no. 188. Retrieved from http://www.visvabharati.ac.in.

Roy, S. (2018, January 09). "Bengal Farmers Smell Opportunity in Aromatic Rice", *The Hindu Businessline*. Retrieved from https://www.thehindubusinessline.com

Roy, Shyamal. (1971). "Profitability of HYV Paddy Cultivation", *Economic and Political Weekly*, 43(26-27), A75-A78. Retrieved from https://www.jstor.org

Rudra, B.C. (2014). "Technological Change and Its Impact on Tenancy Relation in West Bengal", *Economic Affairs*, 60 (1), 1-23.

Saha, A. & Swaminathan, M. (1994). "Agricultural Growth in West Bengal in the 1980s: A Disaggregation by Districts and Crops". *Economic and Political Weekly*, 29(13), A1-A11. Retrieved from https://www.jstor.org/stable/pdf/4400994.pdf?refreqid=excelsior%3A1f09f441462b39340619cd3b7c2b912f

Samanta, G. (2001). "Rural Urban Interaction: A Study of Burdwan Town and Surrounding Rural Areas". 1-289. (Ph.D. Thesis, The University of Burdwan, West Bengal, India). Retrieved from http://hdl.handle.net/10603/66030

Santhy, V., VijayaKumari, P.R., Vishwanathan, A. & Deshmukh, R.K. (n.d). "Legislations for Seed Quality

Regulation in India", *CICR Technical Bulletin No: 38*. Retrieved from http://www.cicr.org.in/pdf/legislation_seed_quality.pdf

Sarkar, A. & Das, A. (2014). "Groundwater Irrigation-Electricity-Crop Diversification Nexus in Punjab: Trends, Turning Points and Policy Initiatives". *Economic and Political Weekly*, 49(52), 64-73.

Sarkar, B. (2017). "The Situation Assessment Surveys: An Evaluation", *Review of Agrarian Studies*, 7(2), 111-122.

Sarkar, P.P. (n.d.). "An Historical Re-reading of Evolving Land-Man Relationship in the Princely State Cooch Behar 1772 -1949", (Ph.D. Thesis, Contextualizing Political Economy of Regional History in Perspective, University of North Bengal, West Bengal, India). Retrieved from http://shodhganga.inflibnet.ac.in/handle/10603/149075

Sarkar, S. & Ghosh, T.K. (2017). "Agricultural Development in West Bengal: An Inter-Temporal Analysis", *Economic Affairs*, 62(3), 483-493.

Satyasai, K.J.S. & Bharti, S. (2016). "Doubling Farmers' Income: Way Forward". *Rural Pulse*, 14, 1-4.

Sekhri S. (2012). "India Policy Forum", Vol. 9, ed. S. Shah et al., Sage, New Delhi.

Sen, A. (1964). "Size of Holdings and Productivity". *The Economic Weekly*, Annual number February, 1964, 323-326. Retrieved from https://www.epw.in/system/files/pdf/1964_16/5-6-7/size_of_holdings_and_productivity.pdf

Sen, A. and Bhatia, M.S. (2004). "State of the Indian Farmer: Cost of Cultivation and Farm Income in India", *Academic Foundation*, New Delhi.

Sharma, H.R. (2007). "Land Distribution and Tenancy among Different Social Groups", *Economic and Political Weekly*, 42(41), 4183-4185. Retrieved from https://opendocs.ids.ac.uk/opendocs/bitstream/handle/123456789/2898/wp260-109273.pdf?sequence=1&isAllowed=y

Sharma, K. (2019). "The Agrarian Distress in India-An Inter-Disciplinary Perspective".

Sharma, V.P. & Thaker, H. (2010). "Fertiliser Subsidies in India: Who Are The Beneficiaries?", *Economic and Political Weekly*, 45(12), 68-76.

Sharma, V.P. & Thaker, H. (2011). "Demand for Fertiliser in India: Determinants and Outlook for 2020", *Indian Institute of Management, Ahmedabad*. Retrieved from http://www.indiaenvironmentportal.org.in/files/demands%20for%20fertiliser.pdf

Shetty, P.K. & Gowda, M.S. (Eds.). (2013). "Innovations in Agricultural Policy". National Institute of Advanced Studies.

"*Shibire dhan becheo taka nei, kshov*" [No Payment after Paddy Selling at Paddy Procurement Centre, Fury]. (2017, June 24). *Anandabazar Patrika*, p.17.

Shiva, Vandana. (2005, Feb 14). "The Indian Seed Act and Patent Act: Sowing the Seeds of Dictatorship", *Grain*. Retrieved from https://www.grain.org/fr/article/entries/2166-india-seed-act-patent-act-sowing-the-seeds-of-dictatorship

Shiva, Vandana. (2013). "The Seeds of Suicide: How Monsanto Destroys Farming", *Global Research*. Retrieved from https://www.globalresearch.ca/the-seeds-of-suicide-how-monsanto-destroys-farming/5329947

Shroff, S. (2019, April 8). "Rethinking India's Battles Against Chronic Agrarian Distress". *The Hindu Centre for Politics and Public Policy*. Retrieved on April 20, 2019, from https://www.thehinducentre.com/the-arena/current-issues/article26754462.ece

"*Shukha dhaner jomi, apomirityu dui chasir*" [Dry Paddy Field, Suicide of Two Farmers]. (2015, May 31). *Anandabazar Patrika*.

Siddiqui, K. (2014). "Growth and Crisis in India's Political Economy from 1991 to 2013", *International Journal of Social and Economic Research*, 4(2), 84-99.

Siddiqui, K. (2015). "Agrarian Crisis and Transformation in India". *Journal of Economics and Political Economy*, 2(1), 3-22.

Siddiqui, K. (2021). "Agriculture, Sustainable Development, and Government Policy in World". *World Financial Review.*

Sidhu, R.S., Singh, S. & Bhullar, A.S. (2011). "Farmers Suicide in Punjab: A Census Study of The Two Most Affected Districts". *Economic and Political Weekly*, 46 (26&27), 131-137.

Siebert, S., Burke, J., Faures, J.M., Frenken, K., Hoogeveen, J., Doll, P., Portmann, F.T. (2010). "Groundwater Use for Irrigation-A Global Inventory". *Hydrol. Earth Syst. Sci.* 14, 1863-1880. https://doi.org/10.5194/hess-14-1863-2010.

Singa, K. & Lokesh, G.B. (2012). "Hulling and Milling Ratio in Major Paddy Growing States: A Case of Karnataka", *Institute for Social and Economic Change*, Bangalore. 1-99. Retrieved from http://www.isec.ac.in13%20Hull%20&%20Mill%20Ratio%20in%20Maj%20Padd%20Grow%20State%20A%2 0Case%20of%20Kar.pdf

Singh J.P. (2006). "Changing Agrarian Relationship in Rural India". *Indian Journal of Agricultural Economics*, 61(1), 36-64.

Singh, K., Singh, S. & Kingra, H.S. (2009). "Agrarian Crisis and Depeasantisation in Punjab: Status of Small/Marginal Farmers Who Left Agriculture". *Indian Journal of Agricultural Economics*, 64 (902-2016-67903).

Singh, S. (2012). "Institutional and Policy Aspect of Punjab Agriculture: A Smallholder Perspective". *Economic and Political Weekly*, 47(4), 51-57.

Singh. & Asokan, S.R. (1988). *Institution Finance in Rural India: Efficiency and Efficacy*, New Delhi, India: Oxford and IBH, 6.

Sinha, S. (2004). "Agriculture Insurance in India: Scope for Participation of Private Insurers". *Economic and Political Weekly*, 2605-2612.

Srinivasalu (2015). "Agrarian Crisis and Farmers' Suicides: Reflecting on the Green Revolution Model", *Centre for Multi-Disciplinary Development Research*, Karnataka, pp. 1-19. Available at www.cmdr.ac.in.

Srinivasan, P.V. (2007). "Agriculture and Food Security". In Shovan Ray (Ed.), *Handbook of Agriculture in India*, New Delhi, India: Oxford University Press, 130-167.

Srinivasulu, K. (2018). "Green Revolution Model and Agrarian Crisis: Towards a Perspectival Critique". *Indian Journal of Public Administration*, 64(2), 193-204.

Srivastav, A.L. (2020). "Chemical Fertilizers and Pesticides: Role in Groundwater Contamination". In *Agrochemicals Detection, Treatment and Remediation* (pp. 143-159). Butterworth-Heinemann.

Srivastava, S.K., Chand, R. & Singh, J. (2017). "Changing Crop Production Cost in India: Input Prices, Substitution and Technological Effects". *Agricultural Economics Research Review*, 30, 171-182.

Swaminathan, M.S. (2016). "National Policy for Farmers: Ten Years Later". *Review of Agrarian Studies*, 6(1), 133-144. Retrieved on August 27, 2018, from http://www.ras.org.in/national_policy_for_farmers_ten_years_later

Swamy, A.V. (2010). "Land and Law in Colonial", Retrieved from https://web.williams.edu/Economics/wp/SwamyLandAnd LawInColonialIndia.pdf

Tejani, S. (2015). "Jobless Growth in India: An Investigation". *Cambridge Journal of Economics*, 40, 843-870. Retrieved from https://www.researchgate.net

Tendulkar, S. (2018, Feb 27). "GM Seed Use has Surged in India, But with No Better Results", *News Click*. Retrieved from https://www.newsclick.in/gm-seed-use-has-surged -india-no-better-results

Thakur, A.K., Ghosh, S. & Mohanty, R.K. (2011). "Saving Water in Rice Cultivation: Technological Options", *Kurukshetra*, 59(4), 30-33.

"The Farmer is Dying and The Country is Rejoicing". (2017, April 6). *Rediff.com*. Retrieved on October 23, 2019, from https://www.rediff.com

Thomas, J.J. (2008). "Financial Sector Reforms and Manufacturing Growth in India: A Preliminary Analysis".

Retrieved from http://www.igidr.ac.in/conf/money/mfc_10/Jayan%20Jose%20Thomas_submission_69.pdf

Thomas, J. & Indira Devi, P. (2015). "Structure and Dynamics of Leased Land Farming in Kerala", 144-150, Retrieved from https://www.researchgate.net/publication/300053351_Structure_and_dynamics_of_lease_land_farming_in_Kerala

Tripathi, A.K. (2013). "Agricultural Price Policy, Output and Farm Profitability, Examining Linkages during Post Reform Period in India". *Asian Journal of Agriculture and Development*, 10(1), 91-111.

Verma, S. (2001). "Impact of Agricultural Mechanization on Production, Productivity, Cropping Intensity, Income Generation and Employment of Labour". *College of Agricultural Engineering, Punjab Agricultural University, Ludhiana*, 133-153.

Verma, S. & Sidhu. M.S. (2009). "Sources, Replacement and Management of Paddy Seed by Farmers in Punjab", *Agricultural Economics Research Review*, 22, 323-328. Retrieved from https://core.ac.uk/download/pdf/6689644.pdf

Vijay, R. (2012). "Structural Retrogression and Rise of 'New Landlords' in Indian Agriculture: An Empirical Exercise", *Economic and Political Weekly*, 47(5), 37-45.

Vijay, R. & Sreenivasulu, Y. (2013). "Agrarian Structure and Land Lease Arrangements-An Investigation in Nine Villages in Andhra Pradesh", *Economic and Political Weekly Supplement*, June 29, 48(26-27), 42-49

Viswanathan , P.K., Thapa, G.B., Routray, J.K. & Ahmed, M.K.(2012). "Agrarian Transition and Emerging Challenges in Asian Agriculture: A Critical Assessment", *Economic and Political Weekly*, 47(4), 41-50.

"Why The Farmers have Stormed Delhi? What They Want?". (2018, Nov 30). *The Economic Times*. Retrieved on December 3, 2018, from https://economictimes.indiatimes.com/news/et-explains/why-the-farmers-have-stormed-delhi-what-they-want/articleshow/66881527.cms

Wikipedia (n.d.). Retrieved from https://en.wikipedia.org/wiki/Stubble_burning

World bank, 2017. Agricultural land (% of land area) |Data [WWW Document]. accessed5.17.20.http://data.worldbank.org/indicator/AG.LND.AGRI.ZS?year_high_desc=false

Yadav, A., Malik, R.K., Bansal, N.K., Gupta, R.K., Singh, S. & Hobbs, P.R. (2002). "Manual for Using Zero-Till Seed-Cum-Fertilizer Drill, and Zero-Till Drill-Cum-Bed Planter. Rice-Wheat Consortium Technical Bulletin Series 4, New Delhi", *India. Rice-Wheat Consortium for the Indo-Gangetic Plains*, 1-22.

Yadu, C.R. & Satheesha, B. (2016). "Agrarian Question in India: Indications from NSSO 70th round". *Economic and Political Weekly*, 51(16), 20-23.

Zaveri, E., Grogan, D.S., Fisher-Vanden, K., Frolking, S., Lammers, R.B., Wrenn, D.H., Prusevich, A. and Nicholas, R.E. (2016). "Invisible Water, Visible Impact: Groundwater Use and Indian Agriculture under Climate Change", *Environ. Res. Lett.* 11 084005

Zaveri, E. and Lobell, D.B. (2019). "The Role of Irrigation in Changing Wheat Yields and Heat Sensitivity in India", *Nat. Commun.* 10 4144

Appendix

Appendix-1.1

Components and Composite Scores for Area Selection

S. No.	CD Block	Percentage of irrigated area to total cultivable area, 2011	Percentage of irrigated area under canal irrigation, 2013-14	Percentage of Deep Tubewell irrigated area to total irrigated area (2013-14)	Yield of aman crop 2012-13	Crop concentration value 2015-16	Share of bargadars to district total (%) 2013-14	Percentage of agricultural labourer to total workers, 2011	Percentage of rural population served by banks (2011)	Composite score (1+2+3+4+5+6+7+8+)
		1	2	3	4	5	6	7	8	9
1	Kalna I	91.92	58.72	20.87	3480.1	0.74	4.08	47.03	18.5	4.07
2	Kalna II	87.15	76.35	21.02	2680.0	0.85	5.03	49.13	14.6	3.84
3	Manteswar	87.72	91.6	5.04	2634.0	1.28	3.72	50.6	20.8	4.35
4	Purbasthali I	72.68	0	47.08	1749.7	0.63	2.62	23.43	24.7	1.7
5	Purbasthali II	96.26	0	50.22	2958.5	0.38	2.28	44.48	27.6	3.38
6	Katwa I	83.8	83.83	14.87	3756.1	1.14	3.80	39.64	28.4	4.72
7	Katwa II	81.71	79.59	16.26	3175.2	1.03	3.13	42.6	19.6	3.88
8	Ketugram I	94.82	0	90.29	3227.4	1.21	4.07	51.57	23.8	5

9	Ketugram II	92.03	0	20.61	3408.6	1.08	1.72	44.05	24.3	3.64
10	Mongalkote	81.1	98.27	1.73	3158.6	1.08	4.62	54.03	34.9	4.99
11	Ausgram I	80.63	99.89	0.11	3099.0	1.00	3.43	63.24	20.7	4.4
12	Ausgram II	65.59	96.86	1.22	3007.8	1.14	4.97	59.19	14.6	3.89
13	Bhatar	88.69	98.28	0.21	3586.3	1.34	6.15	56.67	8.83	4.92
14	Burdwan I	92.69	98.42	0.97	3127.7	1.21	2.80	39.56	25	4.47
15	Burdwan II	89.39	98.06	1.26	2953.1	1.21	3.50	50.2	18.1	4.37
16	Galsi I	92.39	98.23	1.77	3362.8	1.38	4.50	55.13	16.9	5.04
17	Galsi II	99.37	99.43	0.57	3540.0	1.13	2.79	60.69	25.5	5.34
18	Khandaghosh	97.54	94.61	5.39	2948.4	0.81	6.64	55.37	25.1	5.01
19	Jamalpur	78.87	88.09	7.1	2546.9	0.72	7.87	58.38	18.7	4.11
20	Memari I	94.35	97.22	1.99	3288.0	0.83	4.58	52.16	17.4	4.46
21	Memari II	74.16	93.5	5.1	3119.8	0.9	9.79	58.54	20.3	4.77
22	Raina I	92.69	83.82	14.7	3102.4	0.9	4.21	51.83	11	4.1
23	Raina II	99.58	92.83	5.58	3298.3	1.1	3.72	48.66	18	4.71

Source: Author's calculation based on data published by Bureau of Applied Economics and Statistics, Government of West Bengal; District Census Handbook, Barddhaman, 2011; Dey and Mistri, 2018.

Appendix-1.2
Total Variance Explained

Component	Initial Eigen values			Extraction Sums of Squared Loadings			Rotation Sums of Squared Loadings		
	Total	% of Variance	Cumulative %	Total	% of Variance	Cumulative %	Total	% of Variance	Cumulative %
1.	2.262	32.313	32.313	2.262	32.313	32.313	2.120	30.289	30.289
2.	1.416	20.228	52.541	1.416	20.228	52.541	1.457	20.821	51.110
3.	1.205	17.221	69.762	1.205	17.221	69.762	1.306	18.652	69.762
4.	.816	11.652	81.415						
5.	.706	10.085	91.499						
6.	.424	6.056	97.555						
7.	.171	2.445	100.00						

Extraction Method: Principal Component Analysis.

Appendix-1.3
Component Matrix

	Component		
	1	2	3
% of villages having Commercial & Co-operative Bank	-.167	-.332	.768
% of villages having Agricultural Credit Society	.200	.658	.425
% of villages having Agricultural Marketing Society	-.439	.583	.392
% of villages having national high way or state high way with in 5km distance from the village	.511	-.407	.503
% of villages having Pucca Road	.856	.299	.064
% of villages having Mandis/Regular hat with in 5 km	.741	.371	-.154
% of villages having Electrisity for agriculture	.678	-.374	.026

Extraction Method: Principal Component Analysis.
a. 3 components extracted.

Appendix-1.4
Agricultural Infrastructure Index

CD Blocks	Gram Panchayats	Agricultural Infrastructure Index
Galsi-I	Budbud	16.08064165
	Chaktentul	11.07282528
	Loa-Krishnarampur	9.37133241
	Loa-Ramgopalpur	10.10279565
	Mankar	41.03915073
	Paraj	1.825614914
	Potna-Pursa	17.63758245
	Serorai	3.618339944
	Ucchagram	16.24035114
Galsi-II	Adra	10.30253083
	Bhuri	9.849563938

Galsi	22.74203388
Gohagram	1.716657905
Khano	15.34044719
Kurkuba	3.858254141
Masjidpur	5.659162221
Sanko	25.29133374
Satinandi	43.21538611

Source: Author's calculation based on data published by District Census Handbook, Bardhaman, 2011 using PCA.

Appendix-3.1
Recommended Level of Fertilizers for Various Crops in Galsi-I and II CD Blocks of Purba Bardhaman District (per Hectare)

Crops	Basal Stage						Later stages of top dressing				
	N	P	K	N	P	K	N	P	K	Sulphur	Micro nutrients
Aus paddy	70kg	35kg	35kg	30kg	35kg	35kg	40	0	0	10kg	Zinc Sulphate 20-25kg
Aman paddy	70kg	35kg	35kg	18kg	35kg	35kg	1st 35 kg; 2nd 17kg	0	0	20kg	Zinc Sulphate 20-25kg
Boro paddy	130kg	65kg	65kg	32kg	65kg	65kg	1st 64kg; 2nd 32kg			20kg	Zinc Sulphate 20-25kg
Mustard	100kg	50kg	50kg	50kg	50kg	25kg	55kg	0	25kg	30kg	0
Potato	200kg	150kg	150kg	66kg	150kg	50kg	1st 66kg; 2nd 66kg	0	1st 50kg; 2nd 50kg	20kg	Zinc Sulphate/ Borax/Disodium Octaborate

Source: ADA Office of Galsi-I CD Block.

Appendix-3.2
Distribution of Farmer Households in Sample Villages by Manure Use

	Name of Sample Village	No. of farmer households using manure		Name of Sample Village	No. of farmer households using manure
Galsi-I	Kolkol	5	Galsi-II	Kurkuba	5
	Ramgopalpur	4		Sasanga	10
	Jagulipara	4		Satinandi	9
	Bhirsin	12		Sanota	13
	Mollasarul	8		Sankrai	19
	Dwarnari	8		Dayalpur	9
	Chaktentul	10		Taherpur	16
	Mankar	13		Dalimgaria	20
	Bandutia	4		Adrahati	16
	Total	**68**		**Total**	**117**

Source: Field Survey, 2016-2018.

Appendix-4.1
Nominal and Real C2 Cost of Cultivation of Paddy in West Bengal (Rs./hectare)

Year	Nominal C2 Cost	CPI_AL base 2004-05	Real C2 Cost (at 2004-05 prices)
1996-97	16929.63	74.17	22825.44
1997-98	18644.13	77.78	23970.34
1998-99	20945.81	92.49	22646.57
1999-2k	20398.64	90.99	22418.55
2000-01	19031.46	87.69	21703.11
2001-02	21579.37	90.99	23716.20
2002-03	23916.76	91.59	26112.85
2003-04	24109.8	96.4	25010.17
2004-05	24403.45	100	24403.45
2005-06	25162.96	102.7	24501.42
2006-07	26723.2	109.61	24380.26
2007-08	28141.88	118.62	23724.40

2008-09	33046.12	129.73	25473.00
2009-10	38111.55	151.35	25181.07
2010-11	43019.85	168.47	25535.61
2011-12	49142.99	177.78	27642.59
2012-13	55982.79	196.7	28461.00
2013-14	60924.19	219.22	27791.35
2014-15	71840.42	226.73	31685.45
2015-16	72235.44	228.23	31650.28

Source: Computed from data published by Directorate of Economics and Statistics, Ministry of Agriculture and Farmers Welfare.

Appendix-4.2
MSP of Paddy in West Bengal

Year	Nominal MSP	Real MSP (Rs./ Quintal	Year	Nominal MSP	Real MSP (Rs./ Quintal
1996-97	380	512.34	2006-07	580	529.15
1997-98	415	533.56	2007-08	745	628.06
1998-99	440	475.73	2008-09	900	693.75
1999-2k	490	538.52	2009-10	1000	660.72
2000-01	510	581.59	2010-11	1000	593.58
2001-02	530	582.48	2011-12	1080	607.49
2002-03	530	578.67	2012-13	1250	635.49
2003-04	550	570.54	2013-14	1310	597.57
2004-05	560	560	2014-15	1360	599.83
2005-06	570	555.01	2015-16	1410	617.8

Source: Compiled from data published by Directorate of Economics and Statistics, Ministry of Agriculture and Farmers Welfare.

Appendix-5.1

Co-operative Societies of Galsi-I CD Blocks and Break-up of Members as on 31/03/2017

S. No.	Name of Society	Marginal farmer (%)	Small farmer (%)	Agril. Labour (%)	Others (%)	Total Member	Total borrowing members (%)	KCC holder (%)
1	Bandutia SKUS Ltd	44	34.2	20.3	1.5	882	59.2	37.6
2	Budbud SKUS Ltd	42.1	33.9	23.8	0.2	1547	32.3	15.9
3	Chaktentul SKUS Ltd	14.5	57.6	27.8	0	1343	35.4	33.1
4	DVC Agril Labour PACS	6	8.6	85.4	0	753	100	100
5	Galigram SKUS Ltd	29.3	38.9	30.5	1.3	686	34.7	25.7
6	Haripur SKUS Ltd	43.6	35.2	20.7	0.4	724	49	34.5
7	Jagulipara SKUS Ltd	25.8	53.6	18.2	2.4	801	37	39.5
8	Lowa SKUS Ltd	30.9	16.2	52.2	0.7	2658	27.7	18.4
9	Mahunara-Kutruki SKUS Ltd.	24.5	39.7	35.8	0	837	37.6	24.5
10	Mankar SKUS Ltd	23.2	55	21.9	0	915	35.8	5.8
11	Maro SKUS Ltd	33.8	50.5	15.8	0	660	32.6	13.2
12	Monoharsujapur SKUS Ltd	10.5	32.3	57.2	0	1046	33.9	33.3
13	Paraj Paschimpara SKUS Ltd	27.3	50	22.7	0	238	67.2	62.2
14	Paraj Uttarpara SKUS Ltd	24.3	28.1	36.2	11.4	580	50.3	28.3
15	Potna SKUS Ltd	37.6	36.9	25.5	0	822	19.5	18
16	Potna UCACS Ltd	7.9	27	63.1	2	2868	31	12.2

17	Pursha SKUS Ltd	20.1	66.4	8	5.5	473	34.9	31.9
18	Rakona SKUS Ltd	42.6	31	26.4	0	664	29.5	28.9
19	Sandhipur SKUS Ltd	24.7	31.6	43.7	0	776	68.6	57.2
20	Ucchagram Adarsha SKUS Ltd	28.4	42.7	28.9	0	225	26.2	10.2
	Total	**34.7**	**25.4**	**38.8**	**1.1**	**19498**	**38.7**	**27.3**

Source: Indubhusan Chattopadhyay, supervisor of co-operative, Galsi-I.

Note: Co-operative societies against serial number 4 and 12 were noted dormant during of field survey.

Appendix-5.2

Co-operative Societies of Galsi-II CD Blocks and Break-up of Members as on 31/03/13

S. No.	Name of Society	Marginal farmer	Small farmer	Agril. Labour	Others	Total Member	Total borrowing members (%)	KCC holder (%)
1	Babla SKUS Ltd	12.74	18.53	40.15	28.57	1295	29.7	8.11
2	Bahirghanna SKUS Ltd	20.65	65.54	9.24	4.57	920	69.0	50.00
3	Bakta M SKUS Ltd.	29.86	36.29	18.66	15.19	777	49.4	16.09
4	Baramuria SKUS Ltd	22.73	44.95	14.14	18.18	198	55.1	55.05
5	Belan SKUS Ltd	44.54	38.79	7.18	9.48	348	24.4	0.00
6	Bhuri SKUS Ltd	28.23	22.85	32.26	16.67	372	37.1	30.91
7	DalpurSridharpur SKUS Ltd	66.47	14.45	17.92	1.16	173	100.0	98.84
8	GohagramGaramba SKUS Ltd	48.00	19.23	10.00	22.77	650	18.5	0.00
9	Gomai SKUS Ltd	10.71	9.49	67.19	12.61	896	23.8	19.08
10	Hitta SKUS Ltd	43.46	35.64	8.87	12.03	665	66.9	63.31
11	Joy Krishnapur SKUS Ltd	48.35	29.84	11.32	10.49	486	186.0 (?)	37.04
12	Kaitara SKUS Ltd	39.71	25.00	5.29	30.00	340	29.7	0.00
13	Khano Kundupara SKUS Ltd	22.09	47.67	7.44	22.79	430	57.0	53.95
14	Khano Mondalpara SKUS Ltd	46.53	36.81	11.11	5.56	432	51.4	45.60
15	Khanoucacsl Ltd	24.26	39.86	26.34	9.53	577	35.5	32.24
16	Kishorekona SKUS Ltd	43.44	37.28	19.28	0.00	389	68.6	54.24

17	Konerpur SKUS Ltd	47.71	21.10	12.84	18.35	109	22.9	0.00
18	Kurkuba A SKUS Ltd	6.91	47.00	38.15	7.94	1953	14.8	12.03
19	Mashidpur SKUS Ltd	21.67	49.17	11.25	17.92	240	20.8	0.00
20	Mashidpur S SKUS Ltd	27.72	31.77	21.16	19.34	1432	14.0	8.31
21	Mosjidpurucacs Ltd	24.96	24.14	31.91	18.99	1943	18.5	20.07
22	Merual SKUS Ltd	56.60	33.65	6.92	2.83	318	12.6	0.00
23	Sanko new SKUS Ltd	16.13	24.52	25.16	34.19	775	27.2	25.68
24	Sanko SKUS Ltd	42.40	39.22	9.19	9.19	566	14.1	0.00
25	Sarul SKUS Ltd	19.97	47.35	5.69	26.98	756	21.8	19.18
26	Satinandi SKUS Ltd	34.46	41.54	16.00	8.00	325	17.8	0.00
27	Shyamsundarpur SKUS Ltd	33.33	29.89	14.37	22.41	174	54.6	0.00
28	Uro-Chottomuria SKUS Ltd	39.84	33.44	7.20	19.52	625	27.5	11.52
29	Vepur SKUS Ltd	16.22	28.11	0.00	55.68	185	51.4	44.32
	Total	**26.95**	**33.70**	**22.91**	**16.45**	**18349**	**35.3**	**21.39**

Source: Dipak Kr. Chattopadhayay, Supervisor of Co-operatives, Galsi II.

Note: Co-operative societies against serial number 5, 8, 11, 12, 17, 19, 22, 24, 26 and 27 were noted dormant during field survey.

Index

D

G

H

M

N

O

P

Y

Z